The Future of Indian Agriculture

Published by Academic Foundation
in association with the
Institute of Economic Growth, Delhi

सत्यमेव परमो धर्मः

The Future of Indian Agriculture

Technology and Institutions

Editors

Nilabja Ghosh and C.S.C. Sekhar

ACADEMIC FOUNDATION
NEW DELHI

www.academicfoundation.com

First published in 2013
by

ACADEMIC FOUNDATION
4772-73 / 23 Bharat Ram Road, (23 Ansari Road),
Darya Ganj, New Delhi - 110 002 (India).
Phones : 23245001 / 02 / 03 / 04.
Fax : +91-11-23245005.
E-mail : books@academicfoundation.com
www.academicfoundation.com

in association with :

Institute of Economic Growth, Delhi
www.iegindia.org

Disclaimer:
The findings/views/opinions expressed in this book are solely those of the authors and do not necessarily reflect the views of the publisher.

Cataloging in Publication Data--DK
Courtesy: D.K. Agencies (P) Ltd. <docinfo@dkagencies.com>

The future of Indian agriculture : technology and institutions / editors, Nilabja Ghosh and C.S.C. Sekhar.
p. cm.
ISBN 9788171889761
ISBN 817188976X

1. Agriculture--Economic aspects--India--Congresses. 2. Agricultural innovations--India--Congresses. 3. Agriculture and state--India--Congresses. I. Ghosh, Nilabja. II. Sekhar, C. S. C.

DDC 338.10954 23

Typeset by Italics India, New Delhi.
Printed and bound in India.

Printed and bound by The Book Mint, New Delhi.
www.thebookmint.in

Dedicated to
the memory of
Dr Arjun Sengupta

Contents

List of Editors/Contributors

Editors:

Nilabja Ghosh, currently in the faculty of Institute of Economic Growth, Delhi, India, has taught economics and served in reputed institutes as a researcher in agricultural economics and rural development. Her PhD in economics is from the Indian Statistical Institute. She was a consultant with the United Nations University, participated in international and national collaborative research projects and regularly contributes inputs for government policymaking. Her research interests are econometric modelling, agricultural supply related issues, food markets, gender and environment, including climate change issues. She has written several articles for national and international journals and conferences and coauthored two books.

C.S.C. Sekhar is currently Associate Professor at the Institute of Economic Growth, Delhi, India. He holds a PhD degree in economics from the Delhi School of Economics and a bachelor degree in law (LLB) from Delhi University. His main research interests relate to empirical work (econometric modelling) on market structure, price formation, international commodity markets and agricultural trade. His other areas of interest include political economy of development and development economics. He has widely published in journals of national and international repute.

Contributors:

Yogesh Bhatt is a researcher and is currently pursuing PhD research on food security and biofuel production. He is working as a Research Assistant at Institute of Economic Growth, University of Delhi.

Sangeeta Chakravarty works on price volatility and macro models, and on forecasting output in agriculture and national incomes. She is Assistant Professor at Institute of Economic Growth, University of Delhi.

Ramesh Chand is a renowned policy advisor to the Government of India on agriculture. He has worked extensively on food policy, trade and WTO, and agricultural growth and development. At present, he is the Director, National Centre for Agricultural Economics and Policy Research (NCAP), Delhi.

S. Galab is Professor at Centre for Economic and Social Science, Hyderabad, Andhra Pradesh. He has worked on important subjects like effectiveness of public policies related to poverty alleviation including social capital, women empowerment, child labour and livelihoods; and served in committees appointed by the Government, including Farmers Welfare Commission in response to the agrarian crisis, and contributed to policy making.

Danish A. Hashim is presently Director, Economic Policy & Taxation, Confederation of Indian Industry (CII), New Delhi. His research has covered a wide range of sectors including civil aviation, manufacturing, trade, finance and macroeconomics

K.S. Kavi Kumar is Professor, Madras School of Economics, Chennai. He has Masters degree in Engineering from the Indian Institute of Technology-Kharagpur and under his PhD research, he has done "Modelling and Analysis of Climate Change Impacts on Indian Agriculture". His other interests include energy economics and environmental economics.

N. Chandrasekhara Rao is Professor at the Institute of Economic Growth, University of Delhi. Through his publications and other engagements, he has contributed to the understanding of food processing sector, agricultural price policy, farm profitability and food security, and on the impacts of the Bt cotton cultivation on farm incomes and other aspects of agriculture.

P. Prudhvikar Reddy is a Senior Researcher and Field Coordinator in Young Lives India and is associated with Centre for Economic and Social Studies (CESS), Hyderabad. His associations with research institutes, universities and various NGOs have given him rich field experience on Indian agriculture, sanitation, health and rural development. He was involved in longitudinal studies and estimation of cost of cultivation of various crops, returns from agriculture, input use efficiency, poverty and employment in agriculture.

E. Revathi is a Professor at the Centre for Economic and Social Studies (CESS), Hyderabad. Her areas of research are: agriculture economics, gender studies and regional inequalities. She has participated in important policy oriented projects and contributed to issues of agrarian crisis in Andhra Pradesh, farmers and weavers suicides, interventions and their impact on women's access to land

Arjun Sengupta (Late) was a renowned economist. He worked as the Chairman of the National Commission for Enterprises in the Unorganised Sector and as Member of the United Nations Development Program's Commission on Legal Empowerment of the Poor. He was a Member of Parliament of India, representing West Bengal in the Rajya Sabha from 2006 until his death. He was also the Chairman of the Board of Governors of the Institute of Economic Growth till his death.

Nirmal Sengupta is a Visiting Professor, Indira Gandhi Institute of Development Research (IGIDR), Mumbai. Formerly, he was Director, Madras Institute of Development Studies, Chennai. He was educated at the Indian Statistical Institute. He has several published books relating to water and environment.

V. Shunmugam is chief economist, heading the research department at the MCX Stock Exchange. He has a key role in interfacing with regulators, product development & research, risk mapping and product design.

Sukhpal Singh is Professor, Indian Institute of Management (IIM), Ahmedabad, Gujarat. He has written extensively on vertical coordination in agribusiness, especially contract farming and franchising, value chains and small producers.

Vijay S. Vyas is a noted agricultural economist. He has been the Director of IIM, Ahemdabad; Institute of Development Studies (IDS), Jaipur and Senior Advisor, Agriculture and Rural Development Department. Currently, he is Emeritus Professor and Chairman, Governing Board, Institute of Development Studies, Jaipur and a Memeber, Economic Advisory Council to the Prime Minister. He received Padma Bhushan in 2006 from the Government of India.

Foreword

It gives me great pleasure to see this volume reach fruition, an outcome of singular commitment and consistent effort by the editors. As I write this foreword, my mind goes back to the intellectually heady days of 2008 when a series of conferences and seminars on major themes of research at the Institute of Economic Growth was held to mark the Golden Jubilee of the Institute. Agriculture was one of the themes on which we had a two-day conference in which leading experts from different parts of the country and elsewhere participated. A range of themes was discussed and analysed critically; the volume reflects this adequately.

The broad-ranging inaugural and keynote addresses provided fresh insights into longstanding issues, such as, the concern for low growth rates in agriculture in the recent past and for inclusiveness of the process to include small and marginal farmers. In the volume, Chapters 2 and 3, which constitute later versions of these addresses, dwell also on the future of research and development in agriculture and the need to focus attention on the degradation of land and water. Another chapter is devoted to the formation of groups and collective institutions to tackle issues with respect to credit availability and externalities of farming technologies which has been a continuing feature of the Indian scene.

Among other currently significant issues, the nature of markets in agriculture, including futures markets, are addressed in following chapters. Contract farming is examined from different perspectives as well. In the context of agriculture and foreign trade, India's gains from the Doha round of July 2008 are discussed in a subsequent chapter. The volume also gives attention to specialists' views on deficiencies in current engineering-based approaches to water management.

Amongst possible issues for future research, climate change and biotechnology, get their due share of attention in the volume. While econometric methods are employed by the authors to substantiate their hypotheses, the longer period perspective is not lost and readers are introduced to the nitty gritty of the policy aspects as well. Areas of disagreement continue to exist, in particular with regard to biotechnology and the authors reiterate that dissemination of scientific evidence in the public domain will allow more objective debates leading to the much-needed clarity for policymaking. In the same vein, alternative methods of crop-forecasting are evaluated with a comparison of three data-based methods of modelling, each with its own advantages and shortcomings and its simplicity and complexity.

The volume before us constitutes a compelling *tour de force* of the spectrum of issues confronting Indian agriculture from the perspectives of efficiency, equity and sustainability. It is sure to be a valuable resource for students, researchers and professionals for a long time to come.

Kanchan Chopra
Former Director and Professor,
Institute of Economic Growth.

Preface

Completion of 50 years of the Institute of Economic Growth (IEG) was an occasion filled with justifiable pride and a distinct sense of responsibility for the IEG fraternity. A number of academic events were planned to mark the occasion which included distinguished lectures, seminars and conferences among others. Agriculture is one of the disciplines in which IEG has distinguished itself over the last 50 years. It would not be an exaggeration to say that most, if not all, of the eminent agricultural economists of the country were associated with the institute at some point of time or the other. Dharam Narain, C.H. Hanumantha Rao, B.D.Dhawan, S.K. Ray, K. Subba Rao, Kanchan Chopra, Bina Agarwal, Ashok Gulati, Ramesh Chand, Mahendra Dev—have all made rich contributions to the policy discourse both nationally and internationally. The formidable reputation of IEG's agriculture unit is built on the work of these scholars. It is not surprising then that the institute planned a major conference on agriculture, what with the indifferent agricultural growth, agrarian crisis and food security issues facing the country. The conference was held on 23-24 September, 2008 at the Institute of Economic Growth. The present volume is an outcome of the conference proceedings.

The focus on inclusive growth embedded in the Eleventh Five Year Plan has made accelerated agricultural growth and the equitable spread of its benefits imperative for the country. The real GDP in the Indian economy was registering a robust growth of about 9 per cent per annum since 2005-06, but agriculture was growing at less than half of that rate. With about 58 per cent of the workforce still employed in agriculture, this was indeed troubling! Apart from the issue of disparity in intersectoral growth rates, inclusive growth has several dimensions within the agricultural sector itself—equitable growth among different size-groups of farmers, among several regions, across crops etc. The agrarian structure;

input, output and labour markets; investment and capital formation; domestic and international trade; development and adoption of new technology and rural institutions are some of the factors that have a major bearing on agriculture. In fact any policy has to consider all these aspects in a holistic way to be effective. The conference attempted to focus mainly on these issues.

Many people and institutions have helped us in this endeavour. First of all, we would like to thank the Ministry of Agriculture, Government of India for readily agreeing to the idea of the conference and providing the necessary financial support. We thank Prof Kanchan Chopra, the then Director of the institute, for encouraging us and facilitating the conference in every way. We also specially thank the present Director, Prof Manoj Panda for facilitating this present publication. The support of the finance office of our institute is also gratefully acknowledged. When we approached the experts to contribute papers the response was very enthusiastic. We thank all of them for this and for also bearing with us despite the delay in bringing out this volume. We thank the entire administrative staff of the institute who worked tirelessly to make the conference a success. We would like to place on record our sincere thanks to Shailesh Kumar and Yogesh Bhatt for working well beyond the call of their duty, even on weekends, to ensure that everything went without a hitch. M. Rajeshwor's help with the manuscript was extremely valuable. Without their efforts the conference would not have been the success that it eventually turned out to be.

— Nilabja Ghosh and C.S.C. Sekhar

Institute of Economic Growth

1 Introduction

Nilabja Ghosh and C.S.C. Sekhar

The completion of half a century of lifetime of the Institute of Economic Growth (IEG) was an occasion to celebrate and it provided an appropriate forum to deliberate on a number of important issues facing the Indian economy. The future of agriculture was undoubtedly one which aroused considerable interest. This was particularly so in view of the timing of the seminar, when agriculture has, after a relative neglect over many years, suddenly come back to the centre stage of policy concern both in India and elsewhere, thanks to the food crisis that hit the world markets in 2007-08. The contrast sketched by the current unprecedented growth in other sectors of the Indian economy and agriculture's own past success presented a rather gloomy picture of the present agricultural scene in India. With the demon of production-induced food insecurity receding in the aftermath of the green revolution, many of the concerns surrounding agriculture were either assumed to have been resolved or considered to be of little significance. The notable exception to this general perception mostly concerned the domain of ecology that primarily engaged the environmentalists.

Towards the end of the first decade of the new century, several issues have come to the fore. The uncomfortable feeling that production has been in near stagnation for some time, the new and complex paradigm set in by the market liberalisation, greater focus on poverty and food security in a country that was increasingly publicised at the international and national forums for her growing affluence were some of the developments that were beginning to be almost impossible to ignore in the emerging situation. Above all, the alarming tendency clearly visible in food price movements brought back memories of the food shortages faced by the country in the mid-1960s and a reminder that agriculture needs to be accorded serious attention both for the stability of the polity and the sustainability of the economic growth. New technological and institutional possibilities, apprehensions about the effects of global warming as well as the close nexus between agriculture and climate change are additional factors that set the stage for the current discourse. During the one and a half day's deliberations perceptions, analyses and appraisals were conveyed through addresses, panel discussions and research articles by eminent agricultural economists, researchers and other stakeholders. The list of participants included diverse groups such as economists, agronomists, policymakers, academics and development practitioners.

The assortment of presentations can be organised thematically as follows. In the inaugural session, the centrality of agriculture for achieving equity and inclusiveness in development and financing of the sector and the need for workable plans and programmes for the future of agriculture were highlighted. These deliberations emphasised the need for institutional rethinking and an integrated intersectoral outlook to provide incentives and organisational support to promote growth in agriculture with equity. A summary account of the issues and the thoughts surrounding deprivation and non-inclusiveness of development are given in Section II of this introduction. The second theme relates to market reforms in agriculture. Making the agriculture sector competitive by promoting efficiency and by developing innovative institutional measures requires much greater degree of synergy between the market and the state. Section III gives an overview of the issues discussed in the session on markets that include domestic market scenario, emerging institutional mechanisms and international trade negotiations. In Section IV, a third set of questions were grouped together for consideration. Though rather varied in nature, ranging from new technology designed to overcome nature-imposed restraints and adaptation to climate change to improved public food management through crop forecasting, these topics in some way or the other relate to the vulnerability of agriculture to nature and the need to moderate and mitigate the same.

Another, possibly more varied, set of deliberations emerged from the panel discussions in the seminar that added up to a body of perspectives on Indian agriculture at the present juncture. These discussions did seem to revolve around some key concerns but reflected both commonality and variety of impressions. Summarised in Section V, are the views expressed and the discussions that centred broadly around a few appealing ideas for assessing and planning agricultural development

The Way to Agricultural Development

A single issue that seemed to emerge with remarkable frequency in all discussions and comments on agricultural technology and progress was the structure of Indian agriculture in terms of its size distribution. Indeed over 80 per cent of the holdings were small, i.e., less than two hectares in size. Many of the farmers did not even own land, they just leased in small bits of land for farming. This picture, together with the fact that small farms occupied only 40 per cent of the cropped area, draws attention to the inequality that marks Indian agriculture. At the outset the concern seems to be that these small farmers hardly operate at a level that can reasonably earn adequate income that is viable and that they are at a distinct disadvantage in using technology. There may admittedly be some disagreement about this since green revolution already flourished in such a milieu and the farm management data of the 1950s[1] published by the Government of India already showed that small farms in fact were more productive than the large ones. Whether the farm size productivity relation has since reversed, whether the scenario has changed with land fragmentation further bringing down farm size and whether new technologies like the biotechnology and new institutions like group farming can actually address this weakness are the ideas that stimulated a lot of discussion.

A greater concern that seemed to draw the attention of nearly all the speakers who spoke on policy was however the aspect of inclusiveness (although one speaker even expressed some reservation in using the term) and equity in the farming community. The small and marginal farmers were risk-prone, poor and were treated inequitably by the lending institutions. Growth in income of this section of people is very important, regardless of the growth of the upper sections and indeed could be a positive influence on the latter. Financial non-inclusiveness is a blight that Indian agriculture needs to shake off for its development to make any meaningful contribution in this vast country. The poor farmers neither

1. Studies in the Economics of Farm Management in India.

have the necessary access to institutional finance because banks found it costly to lend them nor did the subsidies given on fertilisers and other inputs necessarily reach them because of the poor delivery mechanism.

In the inaugural address, Arjun Sengupta (Chapter 2) expressed his confidence on the sustainability of the high growth rate that is already achieved by the Indian economy at large. On the contrary, the growth rate of the agriculture sector, which is afflicted by resource degradation, was not so impressive (Chapter 3). Far from it, there were signs of stagnation and this meant a poor prospect for reducing poverty in the country. Policy reforms directed towards agricultural growth in fact essentially turn out to be a humanitarian issue though they were much more than that. Vijay S. Vyas (Chapter 3) reflected on important components of public policy, that of land policy, involving both leasing and purchase of land, research and extension policy that have a bearing on the evolution of technology along with the problems of Indian Council of Agricultural Research (ICAR)'s functioning and finally, the price policy with the role that the minimum support price still plays in the economy but the deliberation seemed to come back to the issue of credit and financial inclusiveness. Sengupta did not find growth in agriculture of much meaning if it bypassed the largest section of the population comprising the small and marginal farmers. Finding out a way to direct credit and subsidies to those who matter most remained a serious challenge. While Vyas talked in terms of redefining the so called priority sector for bank finance and the attention that the 'lending size' required in qualifying the definition, Sengupta reflected on providing a public guarantee or reinstating a risk-sharing mechanism to 'incentivise the banks and agents to reach out to the small and marginal farmers in ways that were affordable' to them. He also dwelt on ways of bringing the small farmers together, the merits of cooperatives, on evolving new and more suitable institutions for the purpose and the desirability of an alternative to the special economic zone (SEZ) model as an innovative design for a pro-poor industrial policy.

Galab *et al.* argue (Chapter 4) that although the spate of farmers' suicides points to a crisis in Indian agriculture, much of which is aggravated by policies pursued during the mid-1990s, interestingly it has also resulted in the emergence of collective institutions through the efforts of public policy, corporate sector and civil society initiatives. They documented few success stories in Andhra Pradesh that address the concerns of small farmers and rural women. The paper underlined the fact that collective action as autonomous institutions can be successful even though cooperatives with state support have failed. The state however has an important role to play in empowering these institutions to transact with the market.

Market Reforms

Performance of agricultural markets and policies relating to them are important in their own right. Chapters 5, 6, 7, 8 and 9 discuss the issues related to markets, price formation and policy; domestic and international. Existence of well-functioning input, output and factor markets is a necessary precondition for agricultural growth. The market structure, degree of integration, level of efficiency and quality of institutions—all have a bearing on price formation and welfare gains to producers and consumers. Also, negotiations at the international forums like World Trade Organization (WTO), have major implications for realising the gains from trade in agriculture. This section starts with a review of the state of academic work on agricultural market integration. This is followed by a discussion of alternate and improved channels of marketing such as contract farming. A detailed exposition of the role of futures markets and price discovery mechanism therein is covered next. The section finally concludes with a discussion of the state of WTO negotiations with special focus on the 'July Package'.

In the wake of extensive market liberalisation and economic reforms in many developing countries, market integration studies have assumed renewed importance. Market integration remains a necessary condition, though not sufficient, for Pareto optimality of the spatial competitive equilibrium and hence the interest in empirical testing of market integration. It may be broadly conceptualised as the satisfaction of the Law of One price in some form or the other. Sekhar (Chapter 5) has made a detailed review of the state of literature on market integration with a special focus on India. He points out that the conceptual distinction between market integration and market efficiency is still elusive in the literature. In the Indian context, the techniques adopted to study market integration are still primitive by international standards. The main reason for this is the non-availability of quality basic price data across markets, which needs to be strengthened for meaningful research. Reliable time series data on prices and transaction costs across various important markets is very important to improve the quality of market integration studies in India. Such a dataset can allow application of the recent advances in estimation techniques to study the degree of market integration in India.

Sukhpal Singh (Chapter 6) considered an important market institution that has attracted a lot of attention in the recent discourses on agricultural markets, namely, contract farming. In this chapter, he lays down the basic rationale for contract farming with a lucid explanation of the different models of contract farming, types of contracts and the various pricing practices. He highlights the near total exclusion of small and marginal farmers from contract farming (CF) in India and draws attention to the institutions through which small farmers can coordinate among themselves. Future research in this area can extend by focussing on specific Indian cases of CF and drawing concrete suggestions for policy.

V. Shunmugam and Danish Hashim (Chapter 7) have discussed the aspects of the still fledgling commodity futures markets for agro-commodities in India. They argue that the indirect benefits from futures markets such as institutional changes and improvement in the commodity markets are much more substantial than the direct benefits of higher prices to the farmers. Some of the important advantages they highlight are the transparency that the exchanges brought in respective commodity markets, improvement in the market infrastructure in terms of scientific storage, quality awareness, quality testing and certification, grading and standardisation, scientifically managed warehouses and the warehouse receipts system with the backend tie-ups with the commercial banks. They argue that the futures markets have, with their better penetration into the rural economy and efforts to reach out to the stakeholders, brought the rural economies into the mainstream of economic growth.

The agricultural commodity futures markets have been criticised for their suspected adverse impact on inflationary pressures. The history of the agricultural futures markets in India has been marked by intense regulation and long periods of bans. Even after the opening up of the markets in the wake of trade liberalisation in the 1990s, misgivings continued to generate considerable policy uncertainty. There is little clarity on the question as to whether the futures markets lead to increase in spot prices. Only empirical evidence from scientific studies can firm up the understanding on the impact of the futures markets on spot prices. C.S.C. Sekhar and Danish Hashim attempt to assess the role of futures markets in influencing the spot market prices in three important commodities, namely, wheat, soya and gram (Chapter 8). Using standard econometric methods, they infer that there is little evidence to suggest adverse effect of futures market on spot prices of wheat and soya.

Since the time India became a signatory to the WTO, external trade reforms in agriculture have assumed added importance. Negotiations at the WTO and the implementation of the decisions taken on this forum have profound welfare implications for agricultural producers and consumers. In Chapter 9, Ramesh Chand discusses the July 2008 Package which constitutes one of the important events of the Doha round. He argues that most of the concerns raised by India have been addressed in the July Package as there is a special and differential treatment (S&D) for developing countries in each element of the proposals on Agreement on Agriculture (AoA). He points out that the package is better than the status quo as there is enough flexibility in the package for India to protect its agriculture from import surges. He argues that any further delay in concluding the round would only benefit the developed countries by allowing them to defer reduction of domestic subsidies and lowering tariff peaks.

At the Mercy of the Nature

Unlike most productive activities that take place insulated within concrete and steel, agriculture is largely exposed to the elements of nature. Natural events have always had a say on the performance of agriculture in all countries and societies. Weather is probably the biggest source of threat to crop cultivation among all perils. In India, the farmers are said to be at the mercy of the 'rain-god' even today despite the all-round technological progress in society. The menace of biological pests is another major threat. Not only are the exposures beyond human control, the solutions to the problems, even if available, are not easy. The future of agriculture has to be about how human beings learn to cope with the perils of nature and protect the activity from calamities and shocks.

Nearly in all organised societies, governments have felt it their duty to come to the farmers' aid almost as a routine because of the extraordinary degree of vulnerability that these agents suffer from. The governments' concern invariably creates budgetary pressures and, more disturbingly, sometimes leads to contradictions with other public commitments that involve other sections of the population. Not surprisingly, the vulnerability of agriculture is a major factor in making agriculture an intensely political subject. Moreover, less visibly though indisputably, all sections of population derive benefits from a well-performing agriculture. Any weather induced failure of agriculture can have severe implications for the society in terms of food security, rural poverty, intersectoral migration and inter-temporal incentive effects for food production.

Thus, the uncertainties surrounding crop production are not just the farmers' dilemma but turns out to be a source of considerable unease for the government in its role as a food manager of the country. Indeed climate change portends that temperatures will increase and monsoon may become more wayward in India. Will the monsoon be deficient or will there be a drought? Will the high temperature come in the way of the seed germination or crop maturation? What will be the effect on food production? Will there be untimely rain? Should the government step up its distribution from public stocks and encourage imports? These are some of the questions that disturb the government continually. In addition, devastation by floods requires effective arrangements for rescue and relief. There is a medium and long-run case for building up an efficient system of identifying the possible problems, developing methods of early warning, understanding how farmers cope, creating an effective disaster management system and regulating and using research outcomes to help farmers overcome the natural perils. The emerging concern about the global climate change creates an urgent need for exploration of the possible direction in research to facilitate adaptation to the adverse effects of weather events in agriculture.

Major natural catastrophes may occur once in a while but floods are a recurrent event in India. The Planning Commission estimated that an average of 7.6 million hectares in the country are affected annually by floods of which 3.6 million hectares are cropped area. The Working Group on Water Resources for the Eleventh Plan gives a fairly large estimate of the average annual damage. Loss of human lives and cattle and other social predicaments add to the burden. Financial damage depends on the level of development of the affected area. Moreover damages are higher when the disaster is less expected in the area. A standard strategy is to prevent development of areas liable to disaster. Flood zoning is also a central guideline in India but when large parts of the geographical area of several states is flood-prone and flood occurs repeatedly and even in regions out of bounds, zoning is hardly a practicable proposition. Outright grants and early warning still remain the key methods of managing losses from flood. The structural measures based on engineering development have not been able to convincingly eliminate floods but they can reduce their severities, although this is also arguable. Yet the feasibility of engineering solutions remains unresolved since the colonial times and India's policy too is far from well developed. Nirmal Sengupta (Chapter 10) revisits the issue that still remains a challenge in Indian agriculture and considers the options in the light of new developments.

The spectre of climate change foreboded by scientists and meteorologists has introduced a new dimension to the uncertainty surrounding agriculture as the nature of climatic factors is likely to undergo a transformation exposing farmers to an altogether new and possibly shifting regime. India being a vast country, the range of climatic changes and the adaptation requirements are also likely to be varied. Changes projected in temperature, precipitation and extreme events are likely to have an adverse impact on water resources, agricultural productivity and all those who are involved in that occupation.

In this context it is important to understand whether and how farmers adapt to the changes. Methodologies used for studying the impact of climate change on agriculture however suffer from limitations in their treatment of farmers' adaptation. The 'potential' impact of climate change can be conceptualised by distinguishing among farmers of different attitudinal dispositions. The impact assessments are however vulnerable to the uncertainties associated with the very prediction of climate. K.S. Kavi Kumar (Chapter 11) takes a Ricardian approach to address the subject. The innovative model proposed and implemented by the author not only takes into account the in-season temporal distribution of temperature and rainfall records but interestingly harnesses the powerful methodological advancements made in econometrics to allow spatial autocorrelation in climate sensitivity in order to provide a glimpse of what the future could be like in Indian agriculture.

Despite technological progress, agriculture's performance is restrained by the state of nature. Indeed the green revolution faced criticism for the regional imbalances created because the technology was suitable only to regions that had certain initial advantages. With important lessons learnt about the shortcomings of the green revolution technology, a search for a new source of growth is an obvious response.

In this context the development of biotechnology could be a potential answer. Biotechnology, studies have shown, has the potential to break the current yield barriers, overcoming biotic and abiotic stresses and providing a way of engineering the product to be suitable for the prevailing conditions. The possibilities even surpass what the green revolution achieved. Far from being restrained by natural specificities, it can potentially help plants withstand climatic changes and degradation of water and soil. By developing plant varieties with an in-built resistance to pests, biotechnology can reduce the use of harmful chemical pesticides. Moreover, crops can be bioengineered to incorporate different desirable output traits that can have positive health effects on consumers. The most outstanding instance is that of the Vitamin A-rich 'golden rice' that can potentially combat malnutrition among children in developing countries. Thus, not only production but the potential for poverty reduction appears to be more powerful in biotechnology than it was in the green revolution.

Yet all this does not make the new technology a panacea. The use of the biotechnology is surrounded by controversies and a deep mistrust of its possible and unknown adverse effects. Admittedly, some of the scepticism is based on half-baked and ill-informed ideas. In India, cotton is the only crop in which biotechnology is officially commercialised at the time of the conference. There were many others in the pipeline. Should the floodgates for biotechnology be opened? To answer this there is a need for serious investigation by social and physical scientists based on proven methods and well-informed deliberations and interpretations. Above all, dissemination of scientific evidence in the public domain will allow more objective debates leading to the much-needed clarity for policymaking. N. Chandrasekhara Rao (Chapter 12) is a contribution to the subject that presents the complex issues lucidly along with evidence drawn from rigorous and field level investigations.

The place of public policy in agriculture where markets are expected to provide necessary signals to economic agents is strongly debated. What may perhaps be agreed upon is that the role of the government is likely to change. Paucity of information is known to be a major source of market imperfection. Business procedures need to be informed. Decisions on logistics, trade negotiations, processing all depend on the timely information available on production prospects where production in agriculture is intensely uncertain and complex. Market intelligence will become more important than ever before in the new market-driven regime to enable farmers to plan their sales and storage, and Indian traders and processors to plan their business. Undoubtedly, public intelligence could constitute a most reliable, unbiased and objective support for players in the market. Food supply management will remain to be a prime responsibility of the government and organising the logistics and networks for public purchases, storage and transportation, timely negotiating for imports or exports at favourable world prices and distribution to areas of crisis—all depend on the early availability of timely and reliable forecasts of production. Moreover, even macromanagement and inflation control are linked to agricultural prospects.

Although India has an extensive system of collecting information on crop production, the methods used are not without their limitations. To upgrade the system, there is a need to review the development that has occurred in various disciplines in forecasting techniques and explore the feasibility of making efficient use of such techniques for the purpose. With a number of methods available, conflicting forecast are not surprising. Reconciliation creates a case for scientific assessments also. Such assessments need to compare both costs and benefits of alternate methods and take a view of both qualitative and quantitative attributes of accuracy.

As a simple demonstration, Ghosh *et al.* (Chapter 13) make a comparison of three data-based methods of modelling each with its own advantages and shortcomings and its simplicity and complexity.

Perspectives on Indian Agriculture

Consensus is desirable but pluralism of views can highlight the complexities of issues. A panel discussion organised at the end of the conference provided just that. The panelists, all eminent scholars in their own right and associated with policymaking in several ways, raised unique issues, provided new perspectives (to the vexing old questions) and suggested solutions that had their own distinct hue. Even the discussion that followed displayed the pluralism of thought. There were more disagreements than agreements, yet the meeting left all participants with more to think about.

Progress or Stagnation of Performance?

Growth rate is a key indicator used commonly to measure the performance of agriculture and the robustness of this measure has to be a primary condition. Assessing growth since the times of the green revolution has been a subject of debate on methodological grounds. The panel discussion in the seminar was in some ways a reminder that the resolution of the disagreement was perhaps not complete. The recent performance of Indian agriculture has been dubbed by many as deceleration that creates a challenge for the future. Indeed the growth figures cited as 3.5 per cent for 1981-1997 and 2 per cent in 1996-2005 bear ample testimony to this trend. The situation in Indian agriculture was described as crisis by many among the panelists but a refreshingly different and optimistic perspective was implicit in a suggestion that a small adjustment in the reference periods could say an altogether different story. Indeed the year 2002 was a drought year comparable in its severity only to the year 1987. If this particular single year is omitted from the sample, the growth rate for 2002-2007 would exceed 4 per cent, the much hallowed target growth rate. India's growth rate had moved in consonance with other Asian countries and global price movements. A revival from the collapse seen in 1997-2001 is witnessed all around the world and India is not an exception.

Technology

Technology was a subject that substantially engaged the speakers at the panel discussion. However, perspectives differed remarkably. Lack of a technological breakthrough in Indian agriculture and the current technological 'fatigue' were dominant themes in the discourse though many of the panelists took note of the emergence of biotechnology, particularly the Bt Cotton technology in Gujarat and what it might have in store. The potential of saving on land and water by the use of this emerging technology was highlighted. One panelist felt that an ecological approach to technology was the need of the hour so that a technology for dryland areas and rainfed crops was the key to bringing large gains in agriculture. Moreover, a view was taken that water and soil management could be potential ways to minimise the risk in agriculture when crop insurance was not proving useful. In an apparent departure to this train of thought questions were raised on the conventional means of assessment of a technology. What is technology? Can productivity be the only measure of technology? There is no dearth of instances to show that technology was indeed developing despite constraints such as state-imposed bans on recruitment in research and extension services. The peak food grain production had taken a quantum jump from 216 million tonnes to 230 million tonnes in recent times, rice and wheat yield rates are far higher than they were in the times of green revolution. All this was regardless of the fact that agriculture was gradually moving to lesser quality lands as more fertile lands went into real estate and SEZs. New varieties were developed to grow the same crops in non-traditional conditions. *Channa*, a pulse crop, was earlier grown in North India in the cold climate. Now as they are being replaced by cereals and oilseeds, they have found a place in dry parts of the country as in Andhra Pradesh and Maharashtra. *Tur* dal used to take 360 days to grow in East India but today it takes 90-120 days to mature. Maize has now moved to Andhra Pradesh and Karnataka from the so called BIMARU states, where they are now raised with impressively higher yield rates. Off-season vegetables are a common sight and the yield rates are also higher. Hybrid varieties are developed even for crops known to be difficult cases. Not only biotechnology, but extensive acreage devoted to zero tillage techniques is saving labour and water in agriculture.

Watershed-based farming is serving well and certain ideal conditions in which it works best are identified.

Role of Public Policy

The green revolution that made use of a promising technology in the 1960s and 1970s was predominantly driven by public policy. An enabling environment was provided in which farmers are given access to technology along with the necessary institutions, infrastructure and price incentives—all of which could be compared with the four wheels of a car. The market reforms that ensued in the early 1990s generated an expectation that the state will retreat to a minimal presence and perhaps remain as a safety net provider only, and its main function will be to facilitate smooth operation of the market mechanism and to allow and encourage new market-based institutions to evolve. Inevitably this raises a question as to what role public policy will play in future agricultural development.

Indeed in context of the 'four wheels of the car' metaphor, more has changed since the green revolution days than the well known market *versus* state administration policy mix. When the high yielding technology was introduced in Indian agriculture there was a presence, *albeit* in a spatially and economically confined from, of a class of large farmers who could take up the risk imposed by an unknown technology. Today the share of small and marginal farmers in India who are generally known to be resource poor has increased manifold and agriculture itself has become a much less attractive proposition. New problems have emerged for the producers, an important one being marketing. The market regulations that were once imposed to protect the farmers have outlived their utility. On the other hand vested interest groups have emerged in the marketing chains creating long spreads between what the consumer pays and what the farmer receives. The spread also widens sharply and disproportionately as distances go up. Similar problems arise with the input delivery mechanism, leaving farmers deprived in times of requirement. The price and subsidy policies are recognised to be creating distortions in the market leading to unbalanced use of fertilisers and deficiency of micronutrients in the soil.

Questions also arise over whether the expectations from liberalisation are being fulfilled. The presence of state is difficult to miss even today, procurement continues to be open-ended and the minimum support price (MSP) is being raised with remarkable frequency as the definition of the cost of cultivation to which it is linked is broadened. Is price policy moving towards an income policy? A linkage between administered domestic prices and the international prices is still weak today.

A Group Approach

This is an institutional novelty that seemed to retain its appeal throughout the discussions starting from the inaugural lectures up to the panel discussions. Galab *et al.* in Chapter 4 have documented few success stories in collective action underlining the fact that the collectives need to be autonomous from state control but need to be capacitated by the state at the same time. A more concrete shape was given to this subject in a more focussed way in one of the panelist's address. It has been a puzzle in Indian agriculture why the cooperative movement saw so little achievement and remained so confined in product coverage and geographic spread even after the resounding success of Amul in Western India. At one time cooperative farming was envisioned as the solution to the severe constraints posed by the small size of the farms. The experience with collectivised farming in the communist bloc countries had proved to be a dampener for collective farming. There was very little thinking subsequently on this idea and redesigning its prospects. The only exceptions were the cooperative organisations dealing with agro-services like credit, inputs and marketing. These were however in reality little else than *de facto* parastatal bodies.

The resurgence of the idea of cooperation and group activities is one of the typical features of the post-liberalisation era in India when apprehensions about how the individually small and powerless farmers would face the market where the more resourceful, informed and scale-advantaged commercial entities operate. An obvious answer would be to bring the farmers together to face the market in groups. A few though scattered success stories led to further nurturing of this idea.

The so called group approach to agricultural development proposed in the panel discussion envisaged a great deal of potential in this institution. Organising farmers in groups not just for marketing as in the case of milk in Gujarat but also in production would mean that the farmers would not only gain more bargaining power against the buyers, contracting corporates, the government and the banks but operating together would bring individual farmers greater access to credit, legal support, risk coverage and technology. It would enable greater private investments on productive assets like tubewells which could have been costly for a single small farmer and even enlarge the sown area by doing away with the boundaries between plots.

Special reference was made in this context to women. Farmers at one time used to be predominantly male. With the ravaging socioeconomic changes sweeping the rural environment in the last two decades, things are not the same. Today more than 10 per cent of the rural households are female-headed and may be, 20 per cent are *de facto* so where the males have migrated to cities to earn higher incomes leaving the women to manage the farms. Yet these actual operators of farming rarely possess title to land which puts them at a disadvantage in the market especially in the credit institutions. A radical change in land ownership is required involving distribution of land so that women farmers have adequate access to this vital resource. Even the ownership of homestead land could help women start an enterprise and function effectively.

2 Mainstreaming Small Farms

A New Approach Needed[1]

Arjun Sengupta

The Indian economy is growing at a very high rate. In fact, over the last 10-15 years, our economy has been growing at a much higher rate than probably any other country. The sustainability of the rate of growth is even more of importance, since we had different bouts of economic growth earlier. But since the 1990s, our economic growth seems to have stabilised and we are now, and I accept this claim that with the 1990 reforms, that is, 1992-93 onwards, we have reached a stable trajectory of high income growth. You could call it the baseline; there can be some disagreement on that. But the rate of growth which is a function of the rate of savings, will continue to be very high.

A very crucial point that I want you to appreciate is—the government may not be able to do much to promote growth, but it can do a lot to upset the process of growth. If the government, I am not saying the minimal government, is such that it does not really upset the process of growth, our rate of growth will continue to be at 7-8 per cent minimum, if not 9 per cent which some think is the next target to achieve.

Sustained Growth is not Enough

The point to be realised is that this high rate of growth has bypassed 77 per cent of our population. More than three-fourth of our population have not benefitted from this process of growth. This is the very crucial figure that we should realise. When I mentioned this figure of 77 per cent, Dr Manmohan Singh in the Parliament said that this was the figure that was relevant for 2004-05 that happened to be the NSS Year, beyond which

1. This paper is based on the Inaugural address delivered at the conference.

we do not have the numbers. But that 77 per cent might have come down to 75 per cent, but still an overwhelming proportion of our people have been completely ignored. Anyone who is talking about the future of our economy must be able to say how to do something to affect the 75 per cent positively. This is relevant for agriculture.

I must mention this because there are some figures which are very stark. If you take the small and marginal farmers, because normally the farm size is considered to be less than 0.4 hectares is submarginal; 0.4-1 hectares is marginal; 1-2 hectares is small; above 2 hectares is medium and above 4 hectares is large, it so happens that about 84 per cent of the households belong to small and marginal farmers. The land that they operate upon is 44-45 per cent, the numbers are variable, but the output that they produce is 50 per cent of the output that rest of the economy is producing in agriculture. That means that farmer households above 2 hectares or particularly above 4 hectares are the major output producers or income earners from agriculture. An overwhelming portion of 84 per cent of population do not have adequate livelihood. If we have an agricultural programme which does not affect or which does not clearly plan to change the situation of this 84 per cent of farmers, to my mind, that future of agriculture is not viable. I do not want to get into the notion of viability. You can bring in economics, politics and all kinds of things. It so happens that all the deprived groups—SCs, STs, BCs, OBCs—belong to that particular group. They seem to be all concentrated in a sort of a 'rainbow coalition', as Mayawati calls it. But it is a coalition of the deprived in that particular group.

I do not like the word 'inclusive growth', because I am talking about the growth of these small and marginal farmers, even if that implies that the other farmers are not growing as fast. I do not think that eventuality is likely, because any strategy that will help the small and marginal farmers will help the larger farmers also. The converse is not true and that has been proved. The agricultural growth has picked up. In fact, if the Eleventh Plan numbers are all right, probably the average rate of growth is more than 4 per cent. It has been picking up for the last few years. But I am afraid, it is not happening for the people about whom India should be worried about. I have detailed numbers of state-wise distribution and the state-wise figures, which are different. There are some states in which performances are very bad; there are some states where the numbers are quite different. But the overall picture is that the small and marginal farmers, who are deprived, do not get the benefits of agricultural development.

All of you are aware that there are problems of soil testing; there are problems of soil conservation, technology of production, inputs, seeds, fertilisers and other inputs; there are problems of water supply. After the output is produced, we have the problems of marketing, getting the correct price. All these are problems for both the small and marginal farmers, and the bigger farmers. It so happens that the Government has instituted many programmes to look after these specific problems. They have also been looked after in different programmes. Currently, there is a National Agricultural Policy which has been involved in this.

But unfortunately that again bypasses the small and marginal farmers. They do not have access to proper credit, subsidised inputs, marketing of their products, extension services. The amount of access that they enjoy is far less than the upper sections. Anybody who is looking at the numbers would say that somehow, we have to look at the institutions delivering the services, so that they reach these small and marginal farmers. It is axiomatic and I do not have to talk about it; if you depend on the markets, it will not go to them; market would always choose those who have purchasing power. You have to have specific institutions oriented towards these groups of people.

Incentives and Institutions

When I say this, I am not talking about bringing back the old regime. That did not work either. So, the challenge is to find a method by which whatever growth process that has been created, whatever market access that has been created for the bigger farmers, they are available and accessed by the small and marginal farmers—without upsetting the market forces. But, how can this be done? This is the challenge that we have to face and we need to work on. We must have a programme which would be directed towards the small and marginal farmers. We had a programme in the Fourth Plan, which was given up later and merged with other programmes.

Today we must have a programme which will be directed to this group of people. When I say 'directed', I would mean that you have to work out incentive systems so that the market agents find it worthwhile to channel their services to this particular group of people. This is the fundamental problem that we have. Firstly, the market must be incentivised to put their services in favour of this people; secondly, this group of people should be organised so that they can avail of this particular benefit.

The first part is quite obvious. Many people have worked on how to incentivise the agents. We have found that very little credit is going to the small and marginal farmers; the numbers may be different; but it is hardly 2-3 per cent that is going to agriculture and to this group of people. We have to make credit available to them. How do we do that? We cannot ask the banks to go and lend them. In fact, this has been done. We have been saying, lend without any collateral. Nobody accepts that. Lend to the small and marginal farmers; very few people are accepting that. We have set up a priority sector lending, of which 18 per cent is supposed to go to agriculture, but an overwhelming portion of 18 per cent goes to the bigger farmer. So, we have a priority sector that goes entirely to the upper section of the people. How to change this? You have to do something so that the cost of supplying loan to these people comes down. The only way the cost can come down is when the government or the public sector is willing to share the risk, guarantee the credit. In other words, we look into a totally different banking institution, which is prepared to support this and that banking institution, in turn, will have to be supported by the government.

I am mentioning this because we had a suggestion about creating a special bank for the unorganised sector where all the beneficiaries would be belonging to the unorganised sector. But the important point is to provide them with subsidised credit and to see that the subsidised credit actually goes there. There is no point in subsidising credit when we cannot ensure that it is actually delivered.

A report on banking and financial reform had been produced by Raghuram Rajan.[1] It is one of the finest reports. He has the full section on inclusive banking different from the Rangarajan Committee report[2] which also talks about inclusive banking. It is not merely having a bank branch in

1. *A Hundred Small Steps: Report of the Committee on Financial Sector Reform*. Planning Commission, Government of India, 2008.

2. *Report of the Committee on Financial Inclusion*. January 2008.

my locality but the point is, whether I have access to this credit. Rajan goes into that but he still talks about the conventional banking system and in that conventional banking system, almost without reading this, I could have predicted that he would say, 'remove the cap on interest rates'. A very major recommendation of Rajan is to remove the cap on this. But I am afraid this is not going to help this group of people whom I am talking about, credit will not be available or will not be accessed by the poorest of the poor. Whatever is the rate of interest, it has to be something affordable by the poorest of the poor. You must have a refinancing institution which would be willing to refinance the loans, give to the small and marginal farmers at 7 per cent rate of interest, and the banks would know that this particular amount of money can be refinanced from an organisation.

Banking is not the only problem. Problems are there with seeds with technology, with marketing and with fertilisers too. In fact, our study shows that most of the fertilisers bypass the small and marginal farmers; although there is a substantial amount of subsidy that goes into this input. This is the failure of delivery. The question is also that the small farmers must also be able to absorb the support.

If we have small farmers all over, they have to be forming some kind of an association so that by combination, they can access different services at a market rate or at a subsidised rate if it is available. They can become viable together, while individually they may not be viable. A lot of theories can be built upon that. But what has happened to these kinds of farmers' groups?

The first thing that comes to everybody's mind is cooperatives. The surprising thing about the cooperatives is this—of the people who are availing of any service from cooperatives, the share of those holding up to 1-2 hectares, is hardly 20 per cent. The membership of cooperatives among the small group is only 18 per cent, whereas 50 per cent of members are coming from the upper echelons. The numbers are quite striking. There may be many reasons for this. The first reason could be that our cooperatives are captured by the large farmers and the small and marginal farmers do not have that kind of incentives to get into that. This is however not always true. In Kerala, it will be correct to say this. It has a very viable and a vibrant cooperative movement, but this has not been the case in other places, because we have not been able to ensure that the cooperatives are for the small and the marginal farmers.

So we need to ask ourselves, is it possible to have a viable cooperative system built up with only small and marginal farmers clientele? In other words, what kind of incentives are necessary and what kind of support is necessary in order to work out a programme of this kind to have cooperatives for the small and marginal farmers. A second possibility is the SHG. In Andhra Pradesh there is a new experiment that is going on, which is showing a lot of promise of forming the group of farmers to avail of different services. There are a few different models which are talked about, but many of these are non-farmer organisation groups. Kudumbashree in Kerala is basically a women's organisation, but it is a non-farmers' organisation. Thus, model is there and it can be replicated for farmers.

In West Bengal also, the SHGs have been successful but no concrete study on the SHG experiment has come forth. Therefore, we must have some idea about how to organise the small and marginal farmers so that together they can avail of this opportunity. What kind of SHGs, what kind of groups that can be formed that can be supported by the public sector and what is the method of doing this? We have studied the Andhra experiment; that has just come in and so with the meagre experience in hand we can build up on it.

From Farm to Industry and a New Policy

My final point is very important. Depending only on farm income, however much the farm technology improves, these people will not be able to come out of poverty; when expenditure is more than their income. The number that is available shows that an overwhelming portion of this section is indebted; their expenditures are more than the income. So, they need to have other sources of non-farm income. They will have to get into some kind of industrial promotion. But we need to have a strategy to see that small and tiny units—the small and marginal farmers will only have tiny enterprises—can be formed; those that can be supported and can be developed. This is a new orientation to our industrialisation process. Let the government not involve itself with the industrialisation at large. Big farmers, big industries would be able to look after what they are doing. The

SEZ business is totally unnecessary. They do not need any support from government to form an SEZ. If you want to build these kind of clusters, you have to do this for those who cannot do without support. There is a theory behind what I am submitting to you. We must have a new strategy of development of small and marginal industries and tiny industries that will be there in the rural areas, which will consist of the people coming out from agriculture. This is the new agenda for industrialisation.

Again, the major bottleneck in this proposal is credit. The credit has to go to the units which are employing below 10—actually below 7-8 workers. We have about 58 million units in our country which are employing less than 8-9 people and they get only 2 per cent of the net bank credit. You have to provide some kind of an assistance, either by sharing their risks, by giving credit guarantee or providing them with discounts, refinancing facilities. Whichever way we work it out, we have to see that these small and tiny units can survive and can prosper. Microfinancing has come up today but it has not made much progress, mainly because the amount that is involved is very small in that particular exercise. We need to talk about much larger amount, as it goes on. But this has to be a major plank of our development.

Concluding Remarks

If we are enquiring about the future of Indian agriculture, then we are actually talking about the future of the Indian economy, the economic strategy, economic perspectives and the future of the Indian industry. What kind of industrialisation are we talking about, and what kind of financial system are we actually building up and are they designed in ways so that the vulnerable people get access to the facilities are part of this subject. Finally, we have to think of a new form of planning. We do not need to plan for heavy industries and big industries anymore. We do in fact spend a lot of money and time on big industries and infrastructure. It is still very difficult to understand why we have to spend so much of money to allow private sector to invest in infrastructure. In any economic process, if the economic growth of 7-8 per cent is assured and if you allow foreign investments and all other investments to come, these should be viable on their own. This is the simple textbook economics—investment is the function of the rate of return. This rate of return, in a country, with this high rate of growth is assured; they should come from all over the world. Areas, where without public support, no investment will be forthcoming are those that demand attention. Planning should now be involved with those groups of people fully—the small and marginal farmers, the small and marginal units; small, tiny, microenterprises and the poorest of the poor people who do not have adequate livelihood. This would be a completely different process of planning. So, if we are talking about the future of agriculture, we should talk about all these different aspects of our economy.

3 Policies for Agricultural Development

Issues, Gaps and Contentions

Vijay S. Vyas[1]

I have taken up this theme with two reservations—one, when I am talking about the challenges in the future, I am also looking into the policy aspect, not just the institutional aspect. Secondly, apart from technology which is of course a very important aspect on which I do have a few words to say, I will dwell on credit and also on price policy and how they are impacting growth and delivering the type of growth which we would like to have for the agriculture sector.

We have a whole set of institutions, catering to different aspects of development but these institutions are not congruent to the type of clients that they have to service or at least a major part of the clientele that they have to serve. What I have to say especially on technology is mainly based on our observation made in the Steering Committee (GoI, 2007) for agriculture development in the Eleventh Plan, chaired by Dr Hanumantha Rao, where I had the privilege of being a member. We had certain things to say about the future of agriculture which I thought are still very relevant.

As Dr Sengupta (Chapter 2) said, we are witnessing a very fast rate of growth. It is also very clear that the Indian economy is poised to continue this growth unless and until something very untoward happens. But we also know that while the growth in GDP was spearheaded by industry and services, agriculture was stagnating. As we observed in the Steering Committee report, the decade of 1994-95 to 2003-04, registered the highest growth in GDP and the lowest growth in agriculture. So, the type of growth which we are having and the impact which it is making on different sections are issues of concern.

1. This paper is based on the Key Note address delivered in the conference.

The implication of this is sometimes not properly appreciated. There are many people who say that it does not matter if agriculture is stagnating, we have got a lot of foreign exchange now; the economy is at 8-9 per cent trajectory; the service sector is booming; industry is doing reasonably well; we can import; why should we bother too much about the stagnation in agriculture? It is mainly because unless this sector develops, you will not have the poverty reduction of the type we are talking about.

In the last decade when we had a rate of growth of 8 per cent or so, every year, our rate of reduction in poverty was less than 1 per cent. This is the type of equation which we have to bear in mind when we talk about agriculture growth. Without a growth in the agricultural sector, which is much more participatory than others in spite of all the defects the agrarian structure is known to suffer from, we will not be able to have really any major impact on poverty.

The problems in agriculture have been discussed at length by various scholars and by different committees. Many of us have sat on those committees; many of us have written on these committees. Now these are becoming well known. We have a major problem of degradation of natural resources; the production base of agriculture—this is again not very appreciated—is weakening day by day. Our land resources and water resources are rather deteriorating; they are not developing at all. There is a deceleration of output in all the crops and also in the subsectors, which, at one time, it was suggested will have more dynamism. Subsectors like dairying and fisheries are some of them. You will find that there is a deceleration even here. There is a decline of inputs of fertilisers, and we are not even getting the same return on this input use. There is a declining inefficiency of the input; there is a widening income gap between agriculture and non-agriculture worker, if we were to judge by the GDP per person in agriculture and in industry. More importantly, there is a widening gap between the rainfed agriculture and irrigated agriculture. Rainfed agriculture is not only stagnant, but a deceleration of performance has been observed in the last few years.

A great vulnerability is felt by agriculture today. There is a problem of insecurity, not only because of the production fluctuations, but the vulnerability is increasing also because of the price fluctuations. Most importantly, there is decay in self-help institutions. The cooperative movement, which was, in earlier years, a very vibrant movement, is now a preserve of some of the political groups. They are not delivering to the members at large.

These are the areas of concern that have been expressed by several committees, reports, by scholars working in this area. In the approach also, there is, more or less, a consensus; but action is needed to generate the extra mile; there is need for technology; public investment is needed to create infrastructure, reforms in governance; and existing policies are to be examined and new policies will have to be devised if required.

My paper is mainly centred on the reforms in agriculture policy, but I will also bring in some of the institutional aspects. The policy reforms will have to take into account the present context! There are important features of the present situations that we all need to remind ourselves all the time! What are the initial conditions with which we are starting the growth process? Small and marginal farmers dominate our agriculture in number and also in area; as it has been told by Dr Sengupta, given the state-wise figures, more than 40 per cent of our land is now cultivated in holdings of less than 2 hectares. We cannot ignore 40 per cent of agriculture! So, it is not just a humanitarian concern. It should be humanitarian concern in any case, but it is much more than that. It is also a question of, will we be able to really instill some dynamism in an economy, where 40 per cent of the agricultural land is dormant or which is hardly getting any support?

It is equally important to appreciate that the nature of demand for the agricultural product is changing. It is not what it used to be 10 years back; it is very rapidly changing; we all know the reasons for urbanisation. We also have to reckon with the fact that the regulation of the domestic market for agriculture commodities is weakening. There is privatisation and this privatisation has come in a variety of ways while the state's power of providing a direction and giving some shape to agriculture is weakening.

Finally Indian agriculture, as we all know, is integrating with the world economy. In this context, the policies are very important—the land policy, research and development (R&D) policy, credit policy, marketing policy, trade policy, price policy and policies for risk mitigation. Here, I will be commenting mainly on R&D policy, and a little on credit and price policy.

Policies for Land, Research and Credit

When we talk about the land policy, the policy for reaching out to the small and marginal farmers, there has to be a two-pronged approach to it—one is to enable them to make their holdings viable by leasing or by purchasing land. Both these methods were in operation in our country in 1940s and 1950s, in the first flash of land reforms. But some of that, we had lost. The second way is more important to my mind, which is allowing large farmers to vacate the land and give them incentives to start non-farm activities in the rural areas. Actually, this is happening in all the developing regions of the country, but it needs to be stimulated, deliberately done, because entrepreneurship is not something that anyone and everyone can start. We will need to have a risk-taking ability; we should have access to resources and we should have some knowledge. The large farmers who are now getting less and less interest in direct farming should be encouraged to grow economically and this land should be vacated. But that is a different story.

Coming to agriculture research, we have one of the largest agricultural research systems in the developing world. It is not a newly created system. For ages, we are having this system armed with a very well trained team of scientists, very extensive network of institutions and a tradition of agriculture research of certain type, going back to decades. Yet, when we see the outcome, we will find that this system which is so extensive and which is manned by really very well qualified people, is not able to deliver much. That is the puzzle to which we should now address ourselves.

Before that, let us be very clear about one thing. Whatever we talk in respect of agricultural research and development policy, it has to be basically in the public sector. Though there are attempts to encourage research in the private sector, private sector has been given incentives in terms of tax concessions on the R&D efforts, low duty rates on equipment, open access to public research product, but their role will remain, for quite sometime to come, marginal. The main onus will have to be on the public sector. Why is it so? It is because due to the intellectual property rights (IPRs), even if the private sector gets involved in research, the benefit of it will accrue to only a few. Only those who can afford to take up the research will be able to get advantage of it. It will not be available to the people at large as it is not an open access system.

The second thing that we should also recognise is a change in regime. Earlier we used to get a lot of support from international agriculture research system. We had varieties from Mexico, International Rice Research Institute (IRRI) of the Philippines and rice varieties from all other various international research systems. Now, it is debated whether these institutions themselves are generating any usable research or not. There is a big question mark there, and it is certainly true that there is not much spillover from the international research centres to the NARS, i.e., the National Agricultural Research System in India. We will have to remain really alert to develop our own technologies.

If our public system has not delivered, despite the large network of institutions and trained manpower, there are three classes of reasons for this. One is that the resources allotted to research are very inadequate; two, our R&D policies were not appropriate; and three, our organisation of research and extension activities was faulty. On all of this, many people had commented, in the Steering Committee to which I had made a reference. We had extensively discussed this—the National Commission on Farmers, chaired by Prof Swaminathan had looked into it in much greater detail. But these are the three areas which are now coming as the main concerns as far as development of research in agriculture is concerned.

The amount spent on public research in agriculture is very low compared to total GDP from agriculture; it was a meagre 2.2 per cent during 1980-1985 and only 0.33 per cent in 2004, at a time when China was already spending half a per cent of its GDP from agriculture on research, and since then, the gap has widened. The Steering Committee on Agriculture in the Eleventh Plan, had recommended that at least 1 per cent of agriculture GDP should go into research. I hope that the Government takes it seriously. But the paucity of resource is only one of the reasons; it is not the only reason. The major shortcoming of the research organisation funded by public sectors is that the emphasis has been mainly placed on increasing the yield rates per hectare rather than on improving the productivity of other factors of production. Our land and water resources are deteriorating but our research is not addressing those questions. Our emphasis has been mainly placed on increasing yield. There is little effort in terms of stability of production, which is the key requirement in many parts of the country, especially in the rainfed agriculture. Very little work is done on poor man's crops like pulses or on the other subsectors of agriculture, like horticulture, floriculture, sericulture and animal husbandry. Importance of local agro-ecological factors had not been fully recognised. Very little attention is paid on economising the use of costlier and scarce factors of production such as fertilisers, irrigation, water and farm energy. We are finding that while we are spending more on inputs, we are not getting out production commensurate with investment, and most importantly, little attention is paid to demand—domestic as well as international—for different products, while drawing the research agenda. The concerns are very diverse; we find that they are not meeting the emerging demands in our country.

There are also problems at the organisational level. These are being studied by a group of experts in great detail, specifically by the agriculture scientists themselves in the National Commission on Farmers. Some important facts are coming out from this, as well as the investigations of a number of committees that have looked into ICAR's functioning. They are all unanimous on emphasising some points i.e., there is bureaucratic interference in scientific work, there is lack of coordination among different ICAR institutions and between the State Agricultural Universities. Also, there is lack of coordination between agriculture extension and research and little attention to development of human resources and a lack of involvement of likely beneficiaries from this.

This is the picture today. Let me also say a word on the new area of research, biotechnology. An area of agricultural research that is assuming progressively greater importance the world over is the application of biotechnology for plant and animals. We have had substantial efforts in this area by the private sector. But all these efforts are directed towards improving quality and the output from the seed and also for the high value crops. For the crops which the small and marginal farmers are growing in large areas there is very little that the biotechnology-oriented private sector firms are doing. Progress in public sector is hampered again. Now, at least 3-4 major institutions are there on biotechnology but their progress is hampered because of lack of a comprehensive policy on development; we are still wavering which way we should go, how do we regulate it and how do we encourage certain things. As a result, the contribution from the public sector is not much.

There is however nothing new that I am saying as all these problems have been previously addressed. But unfortunately, there is not much impact of these recommendations and if we really want agriculture to move, then we will have to take these lacunae in investment, in content, and in organisation a little more seriously. The public research institutions will have to be properly funded. Planning of research activities will have to be made more relevant to the needs of the country and the research organisations have to be revamped. Agriculture research should primarily remain in the public domain, with open access for all users.

Let me say a few words on the credit policy. What is our current situation? Bank credit to agricultural sector has increased phenomenally. The Finance Minister wanted to increase agricultural credit by 30 per cent in three years; in two years' time, agricultural credit has been doubled. But the number of rural branches has been reduced and the lending of farmers below Rs 25,000 has declined. So, to whom is this money going? This money is not going to the persons about whom Dr Sengupta is concerned. Many a time, when we talk to the bankers, they say, why do we bother about the bank branches in the rural areas? It is a very difficult case and we can sort them out in various other ways. But because of the absence of bank branches in rural areas, there is segmentation in the credit market. The market which is not served by the banks is served by the private moneylenders or others. We can see the contrast there. So, to say that merely by increasing the bank branches, we will not be able to achieve much, is a partial truth. Certainly we will not. It may not be the only condition, but it is one of the necessary conditions if we want to really reach the people whom we are talking about.

Dependence on non-formal sources of credits has not lessened. We have now got NSSO results on that. Banks' capacity to assist the borrowers by proper advice on farm related issues is woefully inadequate. If we are working in the agriculture sector and do not know what agriculture is, then to serve that group will not be possible. Why this has happened—I will come to it later.

We are also told that credit is not the only requirement, and that we need a lot more than credit, that there should be something called credit plus. What that credit plus is, and what the banks are doing, more than disbursing loans and collecting savings, is not very clear. So, while the concept of credit plus has been recognised, it is conspicuously absent in the actual practice.

Financing Agriculture

Where do we go from there? First and foremost, the way is to change the definition of the priority sector. In the Committee which Dr Sengupta referred to, we suggested that if you are really concerned about the priority sector, take a loan below Rs 25,000 as priority. If some target is to be fixed, it is not much use to say 18 per cent to agriculture, 10 per cent for small farmers, 4 per cent for weaker sector, etc., but we should have a target for priority sector for amounts less than Rs 25,000. It will cover all these various sections that we want to serve basically without making it very untidy.

Another development which is happening is that most of the backward areas—district-wise analysis is now available—are not getting credit adequately. At the state level we now get a better picture. The credit-deposit ratio is better than what it used to be. But if we go to the districts, we still find that the credit deposit ratio is very much adverse as far as the backward districts are concerned. Understandably, it cannot be the same because opportunities for investment are not there; but there should be a limit to this inadequacy. The differences are there, and they should be narrowing instead of widening. Again, there is another very major scheme which we have invented in order to help the banks who are not able to invest in agriculture appropriately—the Rural Infrastructure Development Fund (RIDF). As we all know, this is the fund which is created by the shortfalls which the banks have from the mandated lending of 18 per cent. If the bank is lending only 15 per cent, then it has to give 3 per cent to the pool of RIDF. Now with some of the efforts that some of us made, the default which banks make, is penalised by lower interest rates given to them. But there is now a huge fund for infrastructure creation. But the way in which it is being disbursed needs a very careful examination because it is not serving the purpose basically. The share of indirect financing should be curtailed, and the share of direct financing should be increased. One of the ways by which the targets are met is to lend for bulky amounts such as lending for the purchase of a tractor which is also agricultural lending. But when it comes to direct funding for crop raising or for animal husbandry or fishery, the amounts are declining or at least stagnant.

Then, if we are serious about the credit plus approach, then banks should have a much greater and much stronger relationship with the development institutions. The mechanism is there; we have got district level, state level banking committees. But these are only ornamental; they did not serve any useful purpose. There should be a strong component of agricultural graduates in the banks' recruitment. This is important for the people in the agriculture sector also. Earlier when the banks were nationalised and they were asked to go to rural areas in 1969, there was a huge recruitment of agriculture graduates. In fact, as some of our friends from the agricultural universities will tell us, there was no dearth of job for any good agriculture graduates. Banks were just picking them up. But under the BASEL-I and various other reforms, the recruitment stopped. The earlier batches which were there in the banks went to the so-called mainstream banking. They became executive directors (EDs)—three of whom I know became chairmen of the banks. So, at the branch level, even at the regional level, there is no agricultural expertise. By hammering this again and again at the RBI and at other places, the restriction is now being relaxed and some agriculture graduates are being recruited now. Without some knowledge on agriculture, rural financing cannot be done. Special consideration should be given to farmers' organisations, SHGs, cooperatives, etc. There again, we find that the banks are very reluctant. Very fast growing movement is happening in SHGs. When they try to have some relationship with the banks, banks come out with the argument that the transaction costs are high and they do not have personnel to work out transactions. Clear responsibilities should be identified. These are some ideas which we will have to pursue further when we come to agriculture credit policy.

The Problems with the Price Policy

Let me lastly say a few words on agriculture price policy. In this country, we have an integrated agricultural price policy, which comprises minimum support price (MSP), procurement of grains by public agencies, buffer stocks and public distribution system, each one mutually consistent and mutually reinforcing. This arrangement was made in order to see that the whole system is developed. It served us well for a long time. During the last few years, several distortions have crept into the system; these include extension of coverage of MSP without proper arrangement for procurement; we now have got 28 commodities under MSP, but for only hardly 3-4 commodities, we have got proper arrangement for procurement. The reason why I am mentioning it, is that if we are not able to fulfill the obligations, it only erodes the credibility of the government. Unfortunately, more or less, exclusive emphasis is given on the cost plus approach in determining MSP and enlarging the concept of cost of production. We all can say that we always take into account the emerging demand scenario, international price, etc., but the fact of the matter is that the MSP today is determined by the cost of production. The definition of cost of production is enlarging. Anything that you can think of is brought in without any rhyme or reason. This is making the whole thing more untenable.

To my mind, the worst thing which has happened is the obliteration of the difference between MSP and procurement prices and the conclusion on the role of procurement prices. When earlier we had this integrated system where we had MSP which was assured to every producer to remain in the business—that was the idea—at which the state was prepared to procure any amount which is offered for sale, which it will use for buffer stock or for public distribution. However, the first distortion came when the MSP and the procurement prices became one and the same. Secondly, the role of procurement price was basically to serve as a market-oriented instrument—you procure what you need; you do not procure if you do not need; you procure the amount you need; you procure from the places from where you find it more advantageous. All these various considerations were forgotten; now, we have got only one MSP. So, we do not have any consistent procurement policy. We have open market operation, when there is too much of hue and cry and that is the end of it.

Then we have got the delinking of the trade policy with the price policy, and increasing share of subsidies of the inputs and the outputs and weakening role of the institutions for formulating price policy like the Commission for Agricultural Costs and Prices (CACP). What we need to do now is to emphasise the effective implementation of MSP, strengthen and extend crop insurance, contract farming and forward market. In the long term, MSP is not the solution. But it will take time and we should not hasten the pace; as by doing that, we can really do harm to the people who are vulnerable. Let us move in that direction, but in the meanwhile emphasise an effective implementation of that. Delink MSP and procurement prices—later they should act as an instrument for open market operations to stabilise food grain prices within certain bands.

Decentralisation of the responsibility of procurement and public distribution to states with Centre acting as the lender is another policy of contention. On this again, there was a lot of controversy when we made the suggestion; even the states were not willing to take up the responsibility of procurement and distribution. This is because the other supportive institutions, namely the banks which are advancing food credit to Food Corporation of India (FCI) were not coming forward. Warehousing corporation, which owns a large network of warehousing facilities, was not coming forward too. The states, in any case, were not greatly enthusiastic about it, but they had also other reasons for being more lukewarm about it. However, decentralisation could be useful and in fact, the responsibilities should be given to the Panchayati Raj Institutions (PRIs) for better results.

Concluding Remarks

I will end by saying that it will be an oversimplification if we were to assume that the problems of growth and equity in agriculture sector can be resolved only by state initiatives. However, policy perceived and the programmes implemented by a state have a critical role in addressing these things today and I will say in future also.

References

Government of India (GoI) (2007). *Report of the Steering Committee on Agriculture and Allied Sectors for Formulation of the Eleventh Five Year Plan (2007-2012)*. New Delhi: Planning Commission.

4

Agricultural Distress in Andhra Pradesh

Can Collective Institutions be a Way Out?

S. Galab, E. Revathi and
P. Prudhvikar Reddy[1]

The recent spate of farmers' suicides points to crisis in Indian agriculture and the incidence and magnitude of indebtedness in the rural areas. It has been argued that the present crisis has gathered momentum in the context of agriculture related policy changes in the mid-1990s and its incompatibility with structural changes taking place in agriculture. However, the prolonged crisis situation has also seen revitalisation and the emergence of collective institutions as an answer to the distress situation. These institutions are a result of public policy, corporate sector and civil society initiatives. How far such institutions have proved to be ameliorative is the moot question. This paper details the institutions that have emerged in rainfed areas in Andhra Pradesh as an answer to the crisis and how they have been a way to ease the crisis situation

This paper is organised into four sections. Section I deals with the issues of transformation in agriculture, agriculture related policy reforms and their implications to organising agriculture in Andhra Pradesh. Section II traces the emergence of institutions *viz.*, government, civil society and corporate sector in organising agriculture from the early 2000 period. Summary, conclusions and policy implications of the analysis are presented in Section III.

---

1. The authors acknowledge Prof C.H. Hanumantha Rao and S. Mahendra Dev for their comments and suggestions. They also thank the anonymous referee for the comments and suggestions.

Structural Transformation, Agriculture Related Policy Reforms and Implications

The landholding structure influences the productivity enhancing investments that determine agricultural growth. The institutional changes, market process and mutation of ownership have together brought two remarkable changes in the landholding structure in the state over time, that are: (i) increased proportion of small-marginal holdings accounting for 83 per cent of the agricultural operators and almost half the area under cultivation and (ii) a growing proportion of agricultural labourers. The agrarian economy in Andhra Pradesh as in the rest of India is diverse and built on regional variations.[2] There is a high degree of tenancy accounting for 50 to 60 per cent of the cultivated area in Coastal Andhra much of which is unregistered, without any regulation and with exploitative rents, resulting in deleterious effects on agricultural productivity and growth (Reddy, 2006).

Given the regional disparities in development across the three regions of the state (CESS, 2008), the farmers in dryland areas of Telangana and Rayalaseema made efforts to catch up with the developed districts in coastal Andhra by increasing the proportion of area under non-food crops. Also, groundwater resources are utilised in the dryland areas to catch up with the land productivity in the coastal region. The entire growth in groundwater source of irrigation is due to private investment (Revathi, 2005). This is due to the decline in public investment in irrigation which is manifestation of fiscal reforms. This has in turn ultimately resulted in larger demand for credit to meet fixed capital requirements for private irrigation.

Use of modern inputs like improved seeds, fertilisers and pesticides, is high in Andhra Pradesh, compared to the rest of India. Purchased inputs form the dominant part of the total inputs used indicating increased dependency on the markets for the inputs (Galab and Reddy, 2007). This

2. The state is divided into three regions, Coastal Andhra, Telangana and Rayalaseema based on historical administration and socioeconomic and cultural factors and agro-climatically into five zones: North Coast, South Coast, North Telangana, South Telangana and Rayalaseema.

has led to increased demand for credit to meet working capital needs. Financial liberalisation measures have affected adversely the flow of credit to rural areas. The term lending as well as crop loans by commercial banks to agriculture fell much below the growing credit needs of farmers. The share of small borrowers in the total bank credit declined from 22 per cent in 1991 to 7 per cent in 2001. The share of agricultural credit from the formal institutions (like banks) declined between the pre- and post-liberalisation period in the state of Andhra Pradesh (Reddy and Reddy, 2006). Intrastate development disparities also get reflected in bank disbursals with the more developed Coastal Andhra region cornering a larger share of institutional credit.

Public institutions which were for critical support to farmers have become dysfunctional due to the reformist approach of the state. They include the Andhra Pradesh State Agro-Industries Corporation, Andhra Pradesh State Seeds Development Corporation, Cooperative Sugar Factories, Cooperative Spinning Mills. This is having adverse impact on the farming sector. The fall in the extension support services and decline in the investment in agricultural infrastructure has resulted in decline in the total factor productivity. The above changes can be said to be the manifestations of fiscal reforms (Reddy, 2006).

Trade liberalisation in agriculture has led to destabilisation in the case of several crops. The reduced tariffs on commodities like cotton, spices and edible oil depressed domestic prices. Input markets have been affected, especially the case of seed resulting in high prices, high risk and unfair trading practices in the absence of proper regulatory mechanism. The two major crops after rice, *viz.*, groundnut and cotton have suffered from trade liberalisation. The trade policies compensated for the stagnation of yields in the absence of technological advancement through the tools of 'support price' and 'stability in market prices through trade operations' which influenced growth differentials in major crops in the state. The trend in output prices relative to international prices which signifies outcome of these policies shows that support prices have increased faster in the post-reform period, however, costs have risen more than the rise in prices. On the other hand international prices are volatile, which affect cotton to a greater extent and to some extent groundnut indicating uncertainty of returns. An analysis of the trends in costs per hectare of the crops[3] and output prices clearly show that the costs of all the crops increased faster than the output prices in the state (Galab and Reddy, 2007).

Thus the structural transformation in agriculture, the public policy shift and incompatibility between these two developments created major challenges for organising agriculture in the state in the following ways

- Inadequate credit flows from the formal institutions in the face of rising credit demand to: (i) meet the working capital needs for raising non-food crops by intensive use of modern inputs and (ii) to meet the fixed capital to invest in irrigation infrastructure for improving the land productivity.
- Inadequate regulatory mechanism to check the price and quality of inputs flowing from private sector because fiscal reforms have resulted lower public investment in research and development and extension services.
- Inadequate support mechanisms to track the prices of crops that resulted due to trade reforms.
- Telangana region with a lagged entry into commercial mode of agriculture and Rayalaseema region that has already been into commercial mode of agriculture have suffered relatively more compared to the stable agriculture of coastal regions.

Institutional Arrangements in Organising Agriculture

The institutions that emerge as new revitalised form of existing institutions should address the problems of procurement of inputs at reasonable prices with assurance of quality. Reduction of transaction costs, extension services, cost reducing technologies, MSPs for crop outputs are other requirements, given the existing institutional development in the context of dryland regions of Telangana and Rayalaseema.

---

3. The data of Cost of Cultivation Scheme (CCS) at the state level indicate a steep increase in the costs of cultivation at constant prices with base triennium ending 1983-1984 prices, especially in case of paddy and cotton during the 1990s while there was no such corresponding increase in yield. The increase in cost per hectare indicates the increased use of inputs.

The various institutional arrangements of organising agriculture found in the literature can be typified into four categories, *viz.*, farmer associations, self-help groups (SHGs of farmers and cooperative), contract farming for agro-marketing and agro-processing, input suppliers and consultancy firms (Umali and Schwartz, 1994). The production cooperatives are promoted both by the state and NGOs. The Mutually Aided Cooperative Societies (MACSs) Act passed in 1995 in the state has given flexibility in organising the cooperatives. Women collectives (self-help groups) and farmer collectives (Rythu Mitra groups) are the popular form of collectives promoted by the state[4] and NGOs. Agro-marketing and agro-processing especially contract farming is another institutional arrangement promoted by state and corporate sector. National and multinational corporations are found to be the agencies through which contract farming is implemented.

This section examines the functioning of collective institutions like the cooperatives and also the new initiatives taken by the state in Andhra Pradesh, and community managed institutions and input suppliers in addressing distress factors in agriculture.

State Initiatives

Women's SHGs and their Federations

The Government of Andhra Pradesh (GoAP) has provided a larger space for women's SHGs in its strategy for poverty alleviation and women's empowerment. Women's groups were promoted under Development of Women and Children in Rural Areas (DWCRA) programme. The model of South Asia Poverty Alleviation Programme (SAPAP) was adopted in 1994-1995 which was developed into Andhra Pradesh District Poverty Initiatives Project (APDPIP) from June 2000; and, finally transformed into the Andhra Pradesh Rural Poverty Reduction Project (APRPRP) by 2004-05 also called as 'Velugu' meaning light (Galab and Chandrasekhar, 2003). This programme is called the Indira Kranthi Patham (IKP) in its present form. These are implemented by an autonomous society set up by Government of Andhra Pradesh, the Society for Elimination of Rural Poverty (SERP).

4. Andhra Pradesh has taken initiatives in formation of women's SHGs from the early 1990s.

Social mobilisation is the basis of organising women under IKP. This is done through establishing strong and vibrant people's institutions for the poor in the form of women's SHGs and their federations. SHGs which are federated at the first level are called 'village organisations' (VOs), at second level *mandal samakhyas* (MS) and third level, the district or *zilla samakhyas* (ZS). These institutions target at eradication of poverty, child labour,[5] and forging alliances between these institutions and, the state, the corporate sector and non-governmental organisations (NGOs). The programme enables the women to tackle the conditioning factors that are perpetuating poverty, by supporting demand-driven economic projects to increase and consolidate economic gains on a sustainable basis, through facilitating linkages with credit institutions. More than 90 per cent of the poorest and the poor belonging to scheduled caste and scheduled tribe families in Andhra Pradesh are organised into SHGs. Also a large proportion of the poor SHG members belong to small and marginal farmers' categories deriving livelihood from cultivation and allied activities. Many of them are also victims of rising input costs, crop losses due to pests and depressed prices, especially at the time of harvest (Dev *et al.*, 2007). These organisations have been geared up to address the challenges of organising agriculture through interventions in the market and cost-effective technologies.

Interventions in Input and Output Markets

Interlocking of input and output markets in Andhra Pradesh or elsewhere result in high prices for inputs and low price for output. In order to reduce the transaction costs and to overcome adverse effects of interlinked markets, the IKP started market intervention in 2001 in a small way procuring small quantities of neem seed, subsequently extending to red gram. The scheme of decentralised maize purchase operations with local participation was started in 2005-06.[6] The scheme is managed by the VOs.

---

5. By the end of 2007-08, Andhra Pradesh had 7.08 lakh SHGs organised into 34,269 village organisations, 1,085 *mandal samakhyas* and 22 *zilla samakhyas*.

6. It was decided that village organisations under IKP could procure to a maximum of 4 lakh tonnes at MSP of Rs 540 per quintal in the crop season 2005-06 in seven districts in which the decentralised procurement scheme is in operation through state agencies like MARKFED.

This scheme is implemented in seven districts.[7] At the village level, the VO appoints a committee with five members[8] as the 'Procurement Committee' (PC). The VO gives the information to the farmers about the procurement schedule through interpersonal communication, traditional announcements and local media. The members of the PC will be paid a wage of Rs 50 per day. The infrastructure support for the implementation for the scheme like tarpaulins, weighing machines, moisture meters, sieves, gunny bags for loading and storage facilities are provided by the MARKFED (an institution for marketing of agricultural produce). *Hamalis* provide the labour for collecting the output from the farmers and loading to be sent to the store houses of the MARKFED. In addition, there will be community resource persons who are involved with the procurement activity but more in prior activities like checking the quality, moisture level and giving slips to the farmers for procurement. The IKP gives training to the members of the PC on various aspects of management and carrying the procurement operations. The initial financial requirement for the operation is met by the MSs. The VO is paid a commission of Rs 2.70 per quintal for their services.

A recently concluded study on the assessment of this intervention has revealed: the procurement activity has taken place between 45 to 60 days quite satisfactorily.[9] Most of the VOs have even crossed their target procurement. Awareness about procurement was quite high and the time schedule for procurement of maize from different villages was adhered to. However, the waiting time for procurement was high compared to retailer and hence ways to reduce waiting time needed to be strategised. Before the IKP intervention, the retail trader played an important role accounting for nearly 63 per cent of purchases in the maize market. This institution also provided credit to the farmers for which they sold the produce often at

7. The seven districts in which the scheme is implemented are Karimnagar, Nizamabad, Warangal, Medak, Mahabubnagar, Rangareddy and Adilabad .

8. Usually middle-aged women are placed in the committee as they do not have demanding household responsibilities and can allocate much time for the organisation. An advisory committee is also appointed by the VO consisting of farmers to assist the PC.

9. Most of the farmers got information about the VO purchase scheme directly from VO, IKP and SHG either by attending the meeting or through interpersonal communication.

lower prices and delayed payments. When the farmer sells the crop to the retailer the cost comes to Rs 5.2 per quintal, while it is only Rs 1.6 per quintal when sold to the VO. The gain by selling to VO rather than to the retail trader is more to the marginal farmer compared to the large farmer. On the other hand decentralised procurement has also resulted in gains by shifting marketing from Agricultural Market Committee (AMC) to VO. These gains are due to providing MSP as in the case of AMC and also procuring at doorstep without transaction costs unlike marketing at AMC. Procurement by the VO has especially protected the marginal and small farmers from the retailer network to a large extent. The extent of VO procurement was between 95-100 per cent in all the districts.

Apart from the VOs, the SHGs also played significant role in the procurement activities of the commodity. The incomes of majority of the families of members of SHGs increased as a consequence of the procurement initiatives. However, because of lack of experience, proper estimate of expected maize to be procured could not be made properly in some of the procurement centres. As a result, they were not able to handle the large quantities of maize that arrived. Apart from this, there were no proper godowns, tarpaulins, drying platforms and other infrastructural facilities so as to accommodate large quantities of maize. Further, inadequate supply of gunny bags and inadequate transportation facilities hindered speedy transport of procured grain to nearby AMC centres. Further farmers wanted the VO to enlarge activities to providing inputs like seed, fertiliser and also extend procurement to other commodities. In spite of the above problems, VO procurement centres are considered as huge success (Subrahmanyam *et al.*, 2006).

Apart from marketing, the procurement centres under IKP also made some interventions in inputs markets in the survey districts of Anantapur, Chittoor, Mahabubnagar, Srikakulam and Vizianagaram. For instance in Kotrapalli and Gadduru villages of Chittoor district, members of 14 SHGs have procured groundnut seeds at a lower price.[10] In Savara Jeribhadra, one of the survey villages in Srikakulam, the scheduled tribe members of SHG collectively purchased fertilisers thereby reducing their dependence on

10. It was procured for Rs 1,800 per quintal whereas the market value ranged between Rs 2,200 to Rs 2,300 per quintal.

middlemen, also benefitting due to lower transaction costs and increase in the yields.

IKP interventions were made in achieving value addition to products like lac, soyabean, cotton and red gram in Adilabad; non-timber forest produce (NTFP), cashew in Srikakulam; red gram in Vizianagaram; milk marketing in Anantapur; red gram, castor in Mahabubnagar; NTFP in Chittoor by providing required skills and knowledge especially to the youth (Dev *et al.,* 2004). Reasonable and remunerative prices, regular payments, increased employment opportunities for the poor, reasonable profits, marketing facilities near the villages, near elimination of usurious middlemen and traders, declined interlinked transactions, self-confidence acquired by SHG/VO members to run these units on profitable lines are some of the benefits received by the members and non-members of SHGs. Some constraints like infrastructure gaps, pressure from local traders to withdraw are faced by the groups but they also experienced the potential for value addition of minor forest produce. The survey showed that in all the six districts where market interventions under the APDPIP have taken place, have yielded substantial benefits to the poor and poorest of the poor and also helped to run these units on viable business lines.

The plan is also to have 6,000 procurement centres with at least 1,000 centres in tribal areas in the state in the next five years with adequate infrastructure—storage, drying, value addition facilities to each procurement centre. The vision is very clear in that 'every VO should be working like a mini market yard and every farmer and NTFP collector should not go beyond five kilo meters for marketing their produce' (GoAP, 2009).

Cost-Effective Technology Adoption

As an alternative to excessive use of pesticides and thereby soaring cultivation costs, non-pesticide management (NPM)[11] practices have been started to reduce costs. IKP also started this initiative under the Community Managed Sustainable Agriculture (CMSA) since 2005. It began in a small scale in Mahabubnagar district with 500 acres of red gram

11. Non-pesticide management (NPM) is an alternative model of agriculture largely depending upon replacing external inputs with locally available resources. It utilises farmers' knowledge and skills apart from traditional pest management practices.

cultivation. Farmers could save by reducing expenditure to the extent of Rs 1,500 to Rs 5,000 without affecting production levels much. The highest gain is in the case of chilli and cotton (Rs 7,000-13,000 per acre). This encouraged IKP to launch a pilot of NPM implementation in 25,000 acres in 10 districts with the technical support from an NGO called Centre for Sustainable Agriculture (CSA) and its partners working on resource management in 2005-06.[12] The sole aim was to reduce the cost of cultivation, one of the four objectives of the livelihood framework of IKP *viz.*, increase in income, decrease in expenditure, increase in wage employment opportunities and decrease in risk. The design of the project is based on community management—the VOs and MSs of the IKP.[13]

Initially, CMSA has been coterminous with the village level SHG organisation. Over the years, there have been efforts to build farmer SHGs comprising small-marginal farmers (including women), tenants and other agricultural workers. It is planned that women should comprise half the farmer SHGs, which would mean that many of them will have dual membership both in women SHGs and farmer SHGs. Besides giving representation to women, it would help to build farmers' groups on the organisational strengths of SHGs achieved through years of efforts in capacity building. Presently, credit linkage, procurement, marketing and management of farmers' field schools are all done through the SHGs of IKP.

Apart from considerable reduction in cost of cultivation, there are several other perceived benefits to the community from the above initiative like improved health, reduced dependence on markets for external inputs, improvement in the soil fertility, assured quality of food with better production system, wage employment opportunities through *neem* seed

---

12. This is taken as part of public-private partnership programme between the state government and the IKP.

13. A village activist (VA) is identified for each village—drawn from best practising farmers who coordinates the village level capacity building programmes in the form of farmer field schools. The core of the programme is carried through these schools. A cluster coordinator (CC) is identified for every five villages as a paid activist farmer. At the district level the project is managed by district project manager. There will be state project advisor coordinating at the state level. The MS enters an agreement with an NGO with requisite capacities to provide technical support by way of training, information and campaign. Each district will have one NGO as lead organisation and at the state level the CSA gives the technical support. The financial support is provided by the SERP.

procurement and the preparation of biopesticides, restoration of ecological balance by increasing the population of natural enemies, improvement in the livestock health and reduced pollution of water bodies.

The IKP groups also prepared special purpose microcredit plans (MCPs) for the year 2006 at the village level for accurate assessment of input needs, technical facilitation effort and fund requirement. This plan is prepared based on an assessment of demand from a village for the above mentioned requirements, done through discussion with farmers including tenants. Sufficient time is also given to the farmers for discussions with household members on issues like credit requirement, crop planning and extent of their own resources and other sources of investment received from government. Individual plans thus prepared, will be consolidated at SHG level, and later at VO level. This gives a clear picture on crop planning, budget required, market planning in a village. These MCPs could be submitted to local bankers for raising finances.[14] Village procurement centres also would be established where NPM activity is taken up to ensure farmers get a better price and fair weighing of output.

The IKP intervention, thus started with NPM as an entry point activity, was extended to reducing the usage of fertilisers, to seed management and to village enterprises in the second year. It worked with the Krishi Vigyan Kendras (KVKs) towards zero dependency on external inputs which resulted in above mentioned CMSA in the third year (2005). The NPM intervention started with 450 villages in 10 districts in 2005, scaled up to nearly 1897 villages in 18 districts by 2007-08 covering 6.72 lakh acres (Table 4.1). Government of India has also approved the programme under what is called 'Innovation Window' and accordingly sanctioned Rs 162 crore under Rashtriya Krishi Vikas Yojana (RKVY) for a period of five years from 2007-08 to 2011-12. Thus, CMSA found policy support at both state and national level for scaling up.

14. Banks have been providing credit access to SHGs mobilised under IKP and the latter has also been sensitising the banks to lend for MCPs prepared by the groups.

Table 4.1

Progress of NPM in Andhra Pradesh under Community Managed Sustainable Agriculture

Sl.No.	*Indicator*	*2005-06*	*2006-07*	*2007-08*	*2008-09*
1.	No. of districts covered	10	17	18	18
2.	No. of clusters	90	210	377	628
3.	No. of villages	450	1050	1897	3171
4.	Crop area covered (lakh acres)	0.25	2.00	6.72	13. 81
5.	No. of farmers (in lakh)	0.15	0.80	1.92	3.18

Source: SERP (2008); GoAP (2009).

Besides interventions in input markets other programmes which help the community like the 'food security programme[15] and 'total financial inclusion' (TFI)[16] also were taken up by SHGs under the IKP in the state. The latter scheme is an effort to reduce burden of interest on informal debt on small and marginal farmers by accessing 'soft loan' facility to them and swapping high interest burden sources to low interest institutional sources. Each group is given an amount of Rs 5,00,000 and each member will be entitled to get an amount of Rs 50,000. TFI scheme also provides the members for consumption loans besides loan for swapping of high cost debts from informal sources. This process will be negotiated through the group along with the *gram panchayat* officials in the village.[17] Since this

15. Food security programme called 'Asara' (support) has been undertaken through federations of women's SHGs. Food is advanced to the needy families which can repay on a monthly basis when surplus income is available. The programme which started as 'rice credit line' is expanded to cover several commodities. This programme is largely based on the market procurement by the VOs. In villages which are part of CMSA, the production-procurement-food security programmes are being integrated on pilot basis from 2007-08 onwards.

16. The total financial inclusion (TFI) is being implemented in all states of India. It has been launched by the RBI about a year back. In the state of Andhra Pradesh this scheme is implemented through the IKP. The background for the launching of this scheme is the increasing dependence on non-formal sources of credit which is burdensome due to its high cost.

17. The recipient will be repaying the principal and the interest (at 12 per cent) to the bank through which the member received the amount under TFI. Once in six months the assistant project manager of the DRDA will review the repayment schedules of the individual group members and if repayments are found to be regular then the assistant project officer (APO) will extend the 'Pavala Vaddi' or the 3 per cent interest rate facility to the member.

facility is being extended after appraising the repayment schedule, the interest paid at bank rate (12%) will be refunded to the extent of 9 per cent confining the interest burden to 3 per cent.

Thus the women's collectives provide an organisational base for undertaking community based initiatives like marketing, enabling to receive minimum support price, supply of agricultural input, food security, and reduction in cost of cultivation through alternative farming practices like the NPM enabling women to enter in commodity trading by collectivisation. Impact of these interventions are visible in terms of enhanced price for produce, cash payments, saving on cost of cultivation and transportation, protection from unfair marketing practices, which in all enhanced the bargaining power of producers.

Rythu Mitra Groups

In June 2003, the GoAP launched a programme for the formation of Rythu Mitra Groups (RMGs) consisting of small, marginal and tenant farmers, share croppers and oral lessees in the state. Initially each RMG was provided with Rs 2,500 for capacity building, book keeping and exposure visits. Each group is to be provided a loan of Rs 1 Lakh at subsidised (by state) interest rate of 3 per cent, which the group would lend to the members for investment purposes. The RMGs, with about 15 members in each group, are expected to save and function on the lines of women SHGs. The basic objectives of the RMG approach, in addition to credit are technology transfer, access to market information and market services to farmers. It is also envisaged that group approach would facilitate collective action both in input purchase and output marketing for mutual benefit of members of RMG and make the entire process participatory.

NABARD financed RMGs in the year 2004-05 through a pilot project with the objective of providing access to bank credit to poor farmers with meagre landholdings and tenant farmers taking advantage of a group approach. The pilot project covered 13 districts of Andhra Pradesh, which later was extended to all the districts of the state, as an ongoing programme, in the year 2005-06. The RMGs will be able to access credit right from the day of their formation as seasonality is important in case of crop loans and are eligible for both crop loan and investment credit. While crop loan will be based on area cultivated and scales of finance, the investment credit generally depends on unit cost.[18] For the year 2006-07, it is proposed to provide credit linkage to 57,970 RMG groups. So far 6,531 RMGs have been extended credit with a financial out lay of Rs 80.47 crore.

The approach involving RMGs has immense potential not only as a comprehensive way of delivering various services to poor farmers in a cost-effective way by banking network but also to encourage savings, mutual help to tackle problems collectively and mitigate distress. It is proposed by the state government to link all the possible departmental schemes like seed village programme, integrated scheme of oilseeds, pulses, oilpalm and maize (ISOPOM), farm mechanisation, training programmes, vermi compost, green manuring etc., through RMGs so as to empower them for their sustainability.

An evaluation of the performance of the RMGs shows that they lacked in guidance and adequate training for efficient conduct of group activities. The principal objective of taking up gainful economic activity cannot be fulfilled unless they receive proper training and guidance. Besides, an autonomous umbrella organisation, similar to that of women's SHGs, is needed. A clear policy note indicating the objective, need for restructuring the RMGs, formation of new groups, staff support, training arrangements, supervision arrangement, funding support may be taken up for revamping the RMGs. The comparative analysis of RMGs and farmers SHGs under CMSA shows that group approach would succeed if it is based on proper federal organisational architecture and capacity building (NCEUS, 2008).

Farmers' Cooperatives: New Initiatives

Farmers' cooperatives have been in existence since long time besides those promoted by the Department of Cooperation, GoAP.[19] The

18. During the pilot phase, 4,504 RMGs were extended credit to the tune of Rs 28.95 crore. In the subsequent year i.e., 2005-06, 12,468 RMGs were financed with an amount of Rs 131.77 crore with per group finance at Rs 1.06 lakh benefitting 40,000 tenant farmers. All the agencies *viz.*, CBs (20), DCCBs (9) and RRBs (16) have participated in the programme.

19. The Mulkanoor Co-operative Rural Bank (MCRB) and Marketing Society Limited (MSL) in Bheemdevarapally *mandal* of Karimnagar district established in 1956 falls in this category. The society's services are categorised into financial, marketing, input, consumer, welfare services and community initiatives. The society was initially

contd...

Mulukanoor Cooperative Rural Bank and Marketing Society Limited have been the right institution for programmes meant for farmers from the government and also the recent new technological initiatives like the 'eSagu' and the IFFCO kiosks. These new initiatives help the farmers to access the new technical and market related trends especially related to extension, soil testing and getting correct information at the right time. The society has taken up new initiatives broadly falling under: a) information technology, b) farmers' information and training centre and c) upgraded technology in rice mills. The cooperative society is espousing the spirit of cooperation to enhance collective strength of the people, upgrading it technically and has been a major source of assured and sustainable livelihood in the semi-arid region. In the recent times, the paddy seed developed here also had to face competition from multinational companies like Monsanto (US) which encouraged the cultivation of hybrid paddy seeds. Even though there is attractive commission/salary in the latter case, people still feel that the benefits of a cooperative are not short lived. They offer better safety net, greater certainty, bonus and dividends. The cost and returns are high in the case of hybrid paddy production but also face high risk and instability (Joshi, 2006). Based on the philosophy of Cooperative Development Foundation (CDF), farmers' cooperatives have been formed quite recently as response to the felt collective needs of the disempowered farmers.[20]

...contd...

registered under the Cooperative Act 1964 but later under the Mutually Aided Cooperative Societies Act in 1995. The financial services include lending (loans *viz.*, short, medium and long term) and saving services. The marketing services eliminate middlemen and assure the right price to the member farmers by also providing adequate storage infrastructure for the produce. Processing activity is also taken up in terms of rice mills, cotton ginning mills and processing and marketing of paddy seed which is a value addition and hence fetches a good price for the farmers for proper weighment, grading and free transport. The input services enhance the productivity levels. Credit and non-credit inputs are supplied to the farmers along with technical services. The low cost credit is a big boon for the farmers. Some of the inputs like seed are produced by the society. The general services are rendered under the general and insurance categories where failure of electric motors and crops is insured.

20. One of the NGOs in Warangal district (Grama Vikasa) has promoted a few farmers cooperatives.

Public-Private Partnership (PPP)

'eSagu' Experiment[21] in Providing Timely Extension to Farmers

The main objective of 'eSagu'[22] is to bridge information gap by exploiting developments in information and communication technology. The eSagu is an IT-based personalised agro-advisory system. It aims to improve farm productivity by delivering high quality personalised (farm-specific) agro-expert advice in a timely manner to each farm at the doorsteps without him asking a question. The advice is provided on a regular basis (typically once a week) from sowing to harvesting which reduces the cost of cultivation and increases the farm productivity as well as quality of agri-commodities. In eSagu, the developments in IT such as (database, internet and digital photography) are extended to improve the performance of agricultural extension services. The eSagu system offers next generation agro-advisory tool, and supplements and integrates into the existing agricultural extension system.

In eSagu,[23] rather than visiting the crop in person, the agricultural scientist delivers the expert advice by getting the crop status in the form of digital photographs and other information. The farmers are the end users of the system and can be illiterate. A coordinator is an educated and

---

21. eSagu experiment was started by the faculty of IIIT (Reddy *et al.*, 2005), Hyderabad, firstly in Warangal district.

22. *Sagu* in Telugu means cultivation.

23. The operation of eSagu is as follows. A team of agriculture experts work at the eSagu (main) lab (normally in a city) supported by agricultural information system. One eSagu local centre (few computers and one computer operator) is established for a group of about 10 to 20 villages. Educated and experienced farmers from the villages work as coordinators. Depending on the crop, each coordinator is assigned with a fixed number of farms. The coordinator collects the registration details of the farms under him including soil data, water resources and capital availability and sends the information to the main eSagu system. Everyday, the coordinator visits a fixed number of farms and takes four to five photographs of each farm. A CD is prepared with the photographs and other information and transported to the main system by a regular courier service. The agricultural experts, with diverse background (entomology, pathology, agronomy...) at the eSagu (main) lab analyse the crop situation with respect to soil, weather and other agronomic practices and prepare a farm-specific advice. At the local eSagu centre, the advice is downloaded electronically through a dial-up internet connection. The coordinator collects the advice print out and delivers it to the concerned farmer. In this way each farm gets the proactive advice at regular intervals starting from pre-sowing operations to post-harvest precautions.

experienced farmer who is stationed in the village. Each coordinator is attached to eSagu local centre which contains few computers and a computer operator. Agricultural experts possessing university degree in agriculture provide expert advice. Agricultural information system (AIS) is a computer based information system that contains all the related data. Communication system is a mechanism to transmit information from farms to agricultural experts and vice-versa. When enough bandwidth is not available, photographs from the village to the main system will be transmitted through courier service. However, the advices (text) can be transmitted from the main system to the local centre through dial-up internet connection.

The development of eSagu was started during *kharif* season of 2004. The eSagu system was implemented by delivering advisory to 1,051 cotton farms for the farmers of three villages in Warangal district in Andhra Pradesh. During 2005-06, eSagu system for 4,894 farms has been implemented for cotton, chilli, paddy, groundnut, castor and red gram among other crops totalling 32 farms in 39 villages spread over six districts in Andhra Pradesh. It was also extended to aquaculture from 2005-06 onwards. By 2006-07, the number of crops covered was 27; number of villages covered 147; and number of farms 2,619. Besides advice, the project also provided for an entrepreneurship model by way of providing for inputs and credit.

The key results of eSagu project are as follows. The farmers are satisfied with the expert advice as it is helping the farmers to improve input efficiency by encouraging integrated pest management (IPM) methods, judicious use of pesticides and fertilisers by avoiding their indiscriminate usage. The impact analysis shows that the additional benefit accrued to farmers comes to Rs 3,451 per acre[24] with cost-benefit ratio = 1:3.3; however, cost-benefit ratio to aqua farmers has been higher at 1:7.5. The turnaround time for advice delivery using mobile phones is 24 hours. In some centres, the farmers have paid the nominal subscription fee for the services and they are satisfied with the service.The gains are high for cotton and chilli and gain in yield is the highest than gains in fertiliser and pesticide (Evaluation Report, 2007). This system is encouraging IPM, and

---

24. 1 hectare=2.404 acres.

judicious use of pesticides and fertilisers by avoiding their indiscriminate usage. The other advantages with the data are its usefulness in crop insurance making farm as a unit.

Contract/Corporate Farming

Contract farming can be defined as an agreement between farmers and processing and or marketing firms for the production and supply of agricultural products under forward agreements, frequently at pre-determined prices (Singh, 2000). Contract farming takes care of the processing industry needs. For farmers, negotiating the price risk in the open trade regime became a prime task. In the post-reform period, contract farming has extended to more crops. It is being pushed as a panacea to the current agricultural deceleration and a bailout for the farming community from the unprecedented spate of suicides. Agricultural Produce and Market Committee (APMC) Acts of various states have been amended to suit the needs of the contract/corporate farming. Contract farming may take place in a number of ways suitable to the crops, under various institutional arrangements. Contracts are working well in crops like oil palm and gherkins where it has solved the problem of supply of quality raw material and reduces production risk to the processors and price risk to the farmers (Dev and Rao, 2005). The experience of Punjab Agro Food Corporation shows violation of contract by the company in terms of price payments. The experience of 'large scale advances farm project' popularly known as 'Kuppam' project in contract/corporate farming for the past few years in Andhra Pradesh is a case in point to examine the effects of contract farming. A critical view says that at only a small fraction of cost of supporting the corporate farming by the state, peasant farming can be made viable and vibrant absorbing all modern technological development and meeting the needs of the future (Chowdhry *et al.*, 2002). In the case of food processing companies, the GoAP entered into MoU for which it gives exclusive allotment to set up processing company. However, it is essentially an agreement between unequal parties and caution is required. Some of the important issues that arise are sustainability of crop production and resources, informal nature of contracts without any legal protection, exclusion of small farmers, weak bargaining power of unorganised farmers *vis-à-vis* company. Contract farming in the state is in vogue for the crops of oil palm, cocoa, vegetables, gherkins, marigold, *amla*, baby corn and broiler birds mostly in the districts of East and West Godavari, Chittoor, Anantapur, and around the city of Hyderabad (Viswanadham, 2006).

The issue of cooperative coordination *versus* market coordination comes to the fore, the former gives voice and choice to the farmers to be part of 'service at cost' principle, whereas the latter caters to the corporate interests. One way of doing this is to encourage vigorous bargaining cooperatives or other agricultural producer organisations that can negotiate equitable contracts. NGOs can be involved, both to ensure that the contracts are fair and to provide knowledge inputs. Though the amended APMC Act contains some provisions to regulate contract farming, legal protection for contract growers as a group is essential. This would include setting out clearly what the parties must do and what they cannot do in the areas of delivery, payment, returning goods, price-fixing (Singh, 2005) .

A variant of contract farming, eSafal'[25] documents the experience of an initiative to integrate farmer's production decisions, farm management, extension and marketing, by vertically linking up the agencies that supply the services. eSafal started as a private initiative with the help of a KVK,[26] in May 2005 in six villages in two *mandals* in the district of Karimnagar in Andhra Pradesh.[27] There are five strategic partners in the programme, the farmers' clusters, local KVK, seed company (JK Seeds), pesticides company (Syngenta) and a super-spinning mill from Coimbatore. Farmers are provided with a free soil testing facility with the KVK, and suitable inputs services (seed and pesticide) and farm level extension; farm management for clean cotton supplied by respective partners. Farmers also are advised on integrated pest management techniques. The super-spinning mill is contracted to purchase the cotton at a premium price. All these efforts lead

25. eSafal is an initiative started by the Satyam ICT Group in 2005. The experience of this initiative is documented by a Chelle (2008).

26. KVKs initiated by ICAR in various districts focus primarily on training farmers in newer skills.

27. After conducting few educational camps in each of a dozen villages with PowerPoint presentations combined with visual aids to inform and educate farmers about the internet-linked state-of-art technological facilities and possibilities of new farm management, farmers who volunteer to come forward to join the programme are enlisted on a nominal fee of Rs 100.

to a substantial reduction in cost of production, increased yield and quality and giving value for the product.

The study has obtained field level information from 85 farmers out of 381 farmers who have participated in the programme. Under eSafal programme, ultimately procurement was undertaken from only 100 farmers and rest had to sell their produce in the local *mandi*. The sample farmers are drawn from these 100 farmers. The sample farmers have also sowed Bt and non-Bt seeds in their farms besides participating in the eSafal. This provided case to compare the effectiveness of eSafal controlled farms of the sample farmers with non-controlled farms. The cotton productivity of an average farmer under eSafal is an extra 0.84 qt yield per acre, which is 14 per cent more than the other farms. Cost of cultivation of cotton fell by Rs 1,882 per acre, the additional surplus due to additional yield to value being Rs 4,366, reduced marketing costs at Rs 759 and additional price benefit because of quality enhancement at Rs 1,322 per acre resulting in a gross benefit to the farmer of Rs 8,330 per acre. Thus, new institutions have emerged and few old institutions have revived in response to agrarian crisis in the state. State policy too responded by way of taking up new initiatives and by supporting few community based initiatives. Challenges addressed by different institutions are presented in summary form in Table 4.2.

Table 4.2

Institutions in Organising Agriculture

Sl. No.	*Institution*	*Approach*	*Issues Addressed*					
			Credit	*Inputs*	*Marketing*	*Extension*	*Cost Reduction*	*Food Security*
1.	IKP SHG	Group	√	√	√		√	√
2.	RMG	Group	√	√	√			
3.	Cooperatives	Group	√	√	√	√		
4.	eSagu	Individual				√	√	
5.	Contract farming	Individual/ group		√	√			
6.	eSafal	Individual		√	√	√	√	

Source: Compiled by the authors.

Conclusion

Incompatibility between structural transformation and policies has resulted in unviable agriculture resulting in high levels of indebtedness and even suicides. This is due to the fact that an individual farmer is unable to negotiate with market and the state. It is evident from the analysis that structural transformation coupled with policy shifts has disempowered the farming community.

Promoting collective institutions for the farmers to enable them to grapple with the market and the state is the most effective way forward from the crisis in agriculture. Cooperatives as collective institutions with state support have proved to be a failure, but as autonomous institutions they are a success. The collectives (women or farmers) like Rythu Mitra Groups (RMGs) have proved to be effective community based institutions. The women's SHG, VO network has proved to be an efficient community based organisation stretching beyond thrift activity to undertake larger community issues like sustainable agriculture including a gamut of activities like collective input procurement, collective output procurement and sale, cost reducing technology adoption, getting prices right. All this resulted in the reduction of costs and stable output prices and reasonable profit margins thereby to reduce the indebtedness. Besides they now also became effective medium for any programme like the total financial inclusion, food security making them the hub of community based activity. New experiments have also come up to give a technological solution to the problem like the eSagu. The state however has an important role to play in capacitating the local community to handle challenges. In order to empower the farming community, state has promoted collective institutions of small and marginal farmers. The strengthening of community institutions would enable the farmers to negotiate effectively with institutions of contract and corporate farming too.

References

CESS (2008). *Human Development Report 2007*: Andhra Pradesh. Planning Department. GoAP.

Chelle, Raju (2008). *Vertical Integration in Cotton Farming: A Case Study of e-Safal in Andhra Pradesh*. MPhil Dissertation submitted to the University of Hyderabad.

Chowdhry, K.R., A. Prasada Rao, Venkat and Uma Shankari (2002). "AP Coalition in Defence of Diversity", *A Report on Large Scale Advances Project (Kuppam Project).*

Dev, S. Mahendra *et al.* (2004). *Mid-Term Appraisal of DPIP*. Hyderabad: CESS.

———. (2007). *Mid-Term Appraisal of APRPRP.* Hyderabad: CESS.

Dev, S. Mahendra and N. Chandrasekhar Rao (2005). "Food Processing in Andhra Pradesh: Opportunities and Challenges", *Working Paper* No. 57. Hyderabad: CESS.

Evaluation Report of e-Sagu Experiment in Providing Timely Extension to Farmers' (2007). IIIT, Hyderabad.

Galab, S. and N. Chandrasekhar Rao (2003). "Women and Self-Help Groups: Poverty Alleviation and Empowerment", in C.H.H Rao and S. Mahendra Dev (eds.), *Andhra Pradesh Development: Economic Reforms and Challenges Ahead.* Hyderabad: CESS.

Galab, S. and P. Prudhvikar Reddy (2007). "Agriculture Crisis in Andhra Pradesh: Evidence from the Field". Paper submitted to the Book Project on *Agrarian Crisis in India*, IGIDR.

GoAP (2009). *Annual Report 2008-09*. Department of Rural Development.

Joshi, Sopan (2006). "Know Your Seeds", *Down to Earth.* New Delhi: Centre for Science and Environment.

National Commission for Enterprises in the Unorganised Sector (2008). A *Special Programme for Marginal and Small Farmers.* Report submitted to the GoI.

Reddy, Krishna P. *et al.* (2005). *Evaluation Report of 'E-sagu Experiment in Providing Timely Extension to Farmers.* Hyderabad: IIIT.

Reddy, Narsimha (2006). "Half a Century Travails of Agriculture in Andhra Pradesh", in R.S. Rao *et al.* (eds.), *Fifty Years of AP 1956-2006*. Hyderabad: CDRC.

Reddy, K.S. and P. Prudhvikar Reddy (2006). "Banking Sector Reforms and Equity A sub-Regional Analysis", in R.S. Rao, V. Hanumantha Rao and N. Venugopal (eds.), *Fifty Years of Andhra Pradesh 1956–2006.* Chapter 30. Hyderabad: Centre for Documentation, Research and Communication.

Reddy, V. Ratna and P. Prudhvikar Reddy (2003). "Domestic Price Policies in the Context of Trade Liberalisation", Indian *Journal of Agricultural Economics* 58(3). July-September.

Revathi, E. (2005). "Suicide as Vulnerability: Some Dimensions of Agrarian Distress in Andhra Pradesh". *Mimeo*. Hyderabad: Centre for Economics and Social Studies.

Singh, S. (2000). "Contracting Out Solutions: Political Economy of Contract Farming in the Indian Punjab", *World Development* 30(9): 1621-38 (September, 2002).

Singh, Sukpal (2005). *Contract Farming for Agricultural Development: Review of Theory and Practice, With Special Reference to India*. Centre for Trade and Development.

———. (2000). "Contract Farming for Agricultural Diversification in the Indian Punjab: A Study of Performance and Problems", *Indian Journal of Agricultural Economics* 56 (3): 283-94.

Society for Elimination of Rural Poverty (2008). *Community Managed Organic Farming System in Rainfed Areas, A Report*. Hyderabad: SERP.

Subrahmanyam, S., C.P. Nagi Reddy and R. Nalini (2006). *Maize Procurement by Village Organisations: An Impact Analysis.* Society for Elimination of Rural Poverty, Department of Rural Development, GoAP.

Swamy, Rao (2007). *An Internal Note on Functioning of Cooperatives.* Promoted by Grama Vikasa. Warangal.

Umali, D.L. and L. Schwartz (1994). "Public and Private Agricultural Extension: Beyond Traditional Frontiers", *World Bank Discussion Paper* No. 236. Washington, DC: World Bank.

Viswanadham, N. (2006). *Food and Retail Chains: Case Study of Andhra Pradesh and Punjab.* Institute of South Asian Studies, National University of Singapore.

5 Agricultural Market Integration

A Review of Theory and Empirics with Special Focus on India

C.S.C. Sekhar[1]

In the wake of extensive market liberalisation and economic reforms in many developing countries, market integration studies have assumed renewed importance. Also, the availability of high-frequency time series data on prices at different locations than ever before has enabled researchers to focus intensively on spatial price analysis issues in the recent past. Market integration remains a necessary condition, though not sufficient, for Pareto optimality of the spatial competitive equilibrium and hence the interest in empirical testing of market integration.

Much of the problem in market integration analysis revolves around the concept of 'market integration' itself and the empirical evidence needed to demonstrate that condition. There are essentially two strands of thought. In the first, mainly in the macroeconomics and international economics literature, the conceptualisation of market integration is centred around tradability i.e., transfer of excess demand from one market to another through actual or potential physical flows. Positive trade flows are sufficient to demonstrate that markets are integrated and prices need not equilibrate across markets. It is therefore clear that this concept implies a Pareto inefficient distribution. In the other class of models, competitive price equilibrium exists independent of the trade flows. In the Enke-Samuelson-Takayama-Judge (ESTJ) spatial equilibrium model (Enke, 1951; Samuelson, 1952; Takayama and Judge, 1971), the dispersion of price in two locations for an otherwise identical good is bound from above by cost of arbitrage when trade is unrestricted and, from below when trade quotas

1. The author wishes to thank Mr Santosh Kumar Dash for his very meticulous research assistance. The usual disclaimer applies.

exist. If trade occurs and is unrestricted, then marginal trader earns zero profits and prices in the two markets co-move perfectly.

The other conceptual problem in market integration studies is of the hypothesised market structure. As we will see a little later, there are different categories of empirical models, each based on a different hypothesis. For instance, the method of pairwise correlation coefficients (Lele, 1967; Jones, 1968) is based on the hypothesis that the markets are integrated pairwise. The method of cointegration rests on the assumption of a central market. The other methods emphasise the centrality of transfer costs (Baulch, 1997) in price formation.

We shall first discuss in the following Section the conceptual issues related to market integration and then turn to the empirical issues in assessing the nature and degree of market integration. The paper is organised as follows. In Section II, the theoretical and conceptual issues of market integration are discussed. In Section III, the major empirical approaches to measuring market integration are discussed. In Section IV, a review of market integration studies in India is presented. The final Section concludes.

Market Integration, Market Efficiency and the Law of One Price

The law of one price (LoP) states that if trade occurs between two regions, and if all the arbitrage opportunities are exhausted, then prices are equalised upto the cost of commerce (Barrett, 2001). Market integration may be broadly conceptualised as the satisfaction of LoP in some form or the other. Barrett (2001) argued forcefully on the necessity to make a distinction between market integration and market efficiency. According to Barrett, market integration in micro theoretic terms, represents a Walrasian transfer of excess demand from one market to another, as captured in actual or potential physical flows. In this framework, prices need not equilibrate. This is also the concept of market integration used mainly in the macroeconomics and international economics literature. If integration is equivalent to such tradability, then two markets are not integrated if trade does not occur between them and vice-versa. In this case, existence of integration does not tell us anything about the welfare. On the other hand, market efficiency refers to the satisfaction of zero marginal benefit equilibrium condition. In this sense, it is a statement about welfare—whether there exist potential Pareto improvements in the economy.

The equilibrium concept at the heart of most trade theories is the spatial equilibrium of Enke-Smauelson-Takayama-Judge (ESTJ). In this model, the dispersion of prices in two locations for an identical good is bounded from above by the cost of arbitrage when trade is free and by the trade quotas when trade is restricted. A more generalised version of ESTJ equilibrium condition is:

$$q^{10} = 0 \text{ if } p^0 \le p^1 + \tau^{10}$$

$$q^{10} \in \left(0, q^{-10}\right) \text{if } p^0 = p^1 + \tau^{10}$$

$$\text{and } q^{10} = q^{-10} \text{if } p^0 > p^1 + \tau^{10}$$

where p^0 and p^1 are the prices in markets 0 and 1 respectively, τ^{10} is the cost of transportation from market 1 to market 0, q^{10} is the quantity transported from 1 to 0 and q^{-10} is the maximum trade volume permitted (trade quota) between the two markets. Trade is neither necessary nor sufficient for ESTJ equilibrium. Therefore, it is essential to distinguish between efficiency (price equilibrium) and integration denoted by tradability (Barrett, 2001). This notion of efficiency also implies not only firm level profit maximisation but also a perfectly competitive long-run equilibrium, in which, there is no incentive for entry, because the marginal profits to arbitrage are minimised.

It is notable that there are four variables—prices, transaction costs, trade volumes and trade quotas—underpinning the economic concept of efficiency in the ESTJ model. However, most of the efforts to evaluate market efficiency in the literature focus on just one variable—price, possibly because of paucity of data on other variables.

Although the analysis of markets and prices has been the central theme of economics research for a long time, one class of problems that was difficult to address was the effect of transportation costs on local price and commodity movements when traders were located at many different points. Classical value theory made some special assumptions about locational patterns. In pure competition, it is necessary to assume that: a) all buyers and sellers are similarly located b) all buyers are at one location and all sellers are at some other location or c) even if buyers and sellers are scattered, all the traded units pass through some common node or market.

Enke's (1951) was one of the earliest attempts to solve this equilibrium problem without the rigid assumptions about sellers' and buyers' locations. He used an ingenuous electric circuit to determine equilibrium prices and quantities (traded) when a number of buyers and sellers traded a homogeneous good, subject to two simple conditions—that the freight cost is independent of the volume traded and the trading function is linear. Samuelson (1952), in an influential paper, translated Enke's model into a maximisation problem in linear programming. He then solved the reformulated model to obtain the minimum transport cost. These two papers have spawned vast literature on linear programming models in market analysis.

Spatial market analyses are often carried out by using either the trade volumes or prices but not both. Analyses based on trade flows alone cannot establish whether such trade ensures Pareto efficiency by exhausting all rents to spatial arbitrage or not. Similarly, price-based analyses cannot tell us anything about the actual trade patterns and their role in ensuring equilibrium and efficiency. Barrett and Li (2002) developed a model that uses prices, transfer costs and trade flow data, and estimates a mixture distribution model using the maximum likelihood method combined with a corroborating non-parametric test. They emphasise the need to recognise the conceptual difference between market integration (tradability of products between spatially distinct markets irrespective of efficiency or equilibrium) and competitive market equilibrium (in which all extra-normal profits are exhausted by competitive pressures irrespective of whether trade occurs or not).

To elaborate these concepts a little further, let us assume that the two markets i and j are in competitive equilibrium. Then as per the standard ESTJ model, the marginal profits to arbitrage are zero.

$p_{it} \le \tau_{jit}(p_{it}, p_{jt}, c_{jit}) + p_{jt}$ where p_{it} is the price at location i in time t, τ_{jit} is the transaction cost of spatial arbitrage from location j to location i

in time t. When trade occurs i.e., $\tau_{jit} > 0$, then the equilibrium condition holds with equality and all the rents to spatial arbitrage, defined as $R_{jit} = p_{it} - p_{jt} - \tau_{jit}$, are exhausted (equal to zero). But when trade does not occur, that is when $\tau_{jit} = 0$, then for the efficient equilibrium to exist, $R_{jit} \leq 0$. If trade does not occur even when positive rents exist, that is, $\tau_{jit} = 0$ and $R_{jit} > 0$, then the markets are in disequilibrium.

Summing up, the distinct and yet related concepts of spatial equilibrium and market integration thus rely on three variables; prices, transaction costs and trade volumes.

1) Perfect integration: $R_{jit} = 0$ and $\tau_{jit} \geq 0$

2) Imperfect integration: $R_{jit} \neq 0$ and $\tau_{jit} > 0$

3) Segmented equilibrium: $R_{jit} \leq 0$ and $\tau_{jit} = 0$

4) Segmented disequilibrium: $R_{jit} > 0$ and $\tau_{jit} = 0$

The first case refers to integrated markets with market efficiency while the second case is one of integrated markets without efficiency. The third and fourth cases refer to segmented (non-integrated) markets. The third case reflects efficient equilibrium even though the markets are not integrated while the last case reflects disequlibrium, i.e., potential rents are not realised because of market segmentation.

Empirical Issues in Testing Market Integration

Testing for market integration has evolved over time from the early stages of using bivariate correlation coefficients to the more recent techniques that take into account non-stationarity, common trends and endogeneity of food prices. The literature on market integration can be divided into three broad categories.

i) Early methods based on bivariate correlation coefficients (Lele, 1967; Cummings, 1967; Jones, 1968) and their critiques (Blyn, 1973; Harris, 1979).

ii) Methods accounting for non-stationarity of food prices—Granger causality (Gupta and Mueller, 1982), error correction models

(Ravallion, 1986; Goletti, 1994), Engle-Granger cointegration (Dercon, 1995), Johansen cointegration and common trends (Gonzalez-Riviera and Helfand, 2001; Wilson, 2001; Jha *et al.*, 2005) and panel unit roots (Jayasuriya *et al.*, 2008).

iii) Methods emphasising on transfer costs instead of comovement of prices (Barrett, 1996; Baulch, 1997).

The early work on market integration applied measures of static price correlations or regression coefficients between time series of spot prices at different markets (Lele, 1967; Jones, 1968). Initially, simple bivariate correlation coefficients were used as measures of market integration (Lele, 1967; Cummings, 1967 etc.). Another less common method was of calculating the spatial variance of spot prices and test for its long-run convergence to zero or close to zero (Hurd, 1975). The main problem with these studies is the use of raw prices. Without detrending and deseasonalising the data, the price correlations are normally overestimated (Blyn, 1973).

In an influential paper, Ravallion (1986) has demonstrated a methodology to test for market integration in a dynamic setting. He highlighted the inferential problems with early methods using correlations and simple regressions and proposed an alternate multivariate model of market integration within which various market integration conditions can be formulated and tested as nested hypotheses. This model is the first major attempt to study market integration in a systematic manner. The model is a dynamic formulation of the following static price formation model among N markets where market 1 is the central market.

$$p_1=f(p_2, \ldots\ldots\ldots\ldots, P_N, X_1) \quad \ldots(1)$$

$$p_i=f(p_1, X_i) \quad \ldots(2)$$

where X_i(i=1……N) is a vector of other exogeneous variables influencing the ith market.

The dynamic reformulation of the model for T periods and N variables is as follows.

$$p_{1t}=\sum_{j=1}^{n} a_{1j}p_{1,t-j}+\sum_{k=2}^{N}\sum_{j=0}^{n} b_{1j}^{k}p_{k,t-j}+X_{1t}c_1+e_{1t} \quad \ldots(3)$$

$$p_{it} = \sum_{j=1}^{n} a_{ij} p_{i,t-j} + \sum_{j=0}^{n} b_{ij} p_{1,t-j} + X_{it} c_i + e_{it} \qquad ...(4)$$

where a's, b's and c's are the parameters to be estimated and e's are the error terms. A wide range of possible hypotheses can be formulated and tested as parameter restrictions on Equation (4).

i) Market segmentation: $b_{ij}=0$ for j=0,............,n.

ii) Short-run market integration: A price increase in the central market is immediately transmitted to the i th market if $b_{i0}=1$.

a) Weak short-run integration: When lagged effects are present, that is, if either $\sum_{j=1}^{n} a_{ij} \neq 0$ or $\sum_{j=1}^{n} b_{ij} \neq 0$

b) Strong short-run integration: No lagged effects if $\sum_{j=1}^{n} a_{ij} = \sum_{j=1}^{n} b_{ij} = 0$. In such cases the i th market is integrated with the central market within one time period.

iii) Long-run market integration: A long-run equilibrium is one in which market prices are constant over time, undisturbed by the local stochastic effects. In this case, Equation (4) takes the form $p_{it} = p_i^*$, $p_{1t} = p_1^*$ and all $e_{it}=0$. Then Equation (4) reduces to:

$$\frac{p_i^* = p_1^* \sum_{j=0}^{n} b_{ij} + X_{it} c_i}{1 - \sum_{j=1}^{n} a_{ij}} \qquad ...(5)$$

Long-run market integration now requires that the coefficient of:

$$p_1^* = 1 \Rightarrow \frac{\sum_{j=0}^{n} b_{ij}}{1 - \sum_{j=1}^{n} a_{ij}} = 1$$

$$\Rightarrow \sum_{j=1}^{n} a_{ij} + \sum_{j=0}^{n} b_{ij} = 1 \qquad ...(6)$$

This is the condition for long-run market integration.

This implies that if the conditions for short-run integration described by (ii b) which is, $b_{i0}=1$ and $\sum_{j=1}^{n} a_{ij} = \sum_{j=1}^{n} b_{ij} = 0$ hold, then the above long-run integration condition (Equation 6) is satisfied but the reverse is not true.

Vast literature has emerged using this model combined with time series techniques (Alexander and Wyeth, 1994; Palaskas and Harriss-White, 1993; Dercon, 1995; Behura and Pradhan, 1998; Kumar and Sharma, 2003; Harris, 1979; Ardeni, 1989; Goodwin and Schroeder, 1991). Methods like cointegration and error correction have gained a great degree of popularity. They broadly test for the following: i) existence of long-run relationship among prices in different markets, ii) existence of unidirectional and bi-directional causality and iii) dynamics of short-run adjustment to deviations from long-run equilibrium.

Alexander and Wyeth (1994) used the Granger representation theorem to recast the Ravallion model into an error correction model. But their test of bidirectional causality (from central to peripheral markets and vice-versa) is flawed as they mix variables in levels p_{t-1} and R_{t-1} that are non-stationary and variables in first differences that are stationary. Their justification to Dercon's (1995) criticism on this point, that an unrestricted vector autoregression (VAR) has been used to maintain consistency with Ravallion model is unconvincing as such an approach makes the entire estimation process open to question. The use of classical F tests, rather than the LM and LR tests, is also questionable. Dercon (1995), attempts to improve the model by introducing dummies for liberalisation and peace time periods and thereby allows for changing market structure rather than testing for long-run integration of a static structure. However, his results also suffer from the use of F-tests mentioned in case of A-W study. Other important studies that attempted to extend Ravllion model using Johansen cointegration and other time series methods are Gonzalez-Riviera and

Helfand (2001), Wilson (2001) and Yavapolkul *et al.* (2004). In these studies, spatial price analysis has been carried out by testing for long-term cointegration and short-run error correction mechanism by using price data at different geographic locations. Some of the underlying assumptions of the model are the linear pricing and stationary transaction costs.

Despite their popularity, these methods suffer from a fundamental flaw. These approaches rely on several strong and simplifying assumptions. Tests of hypothesis of market efficiency are indistinguishable from those of the underlying assumptions of the model. For instance, in the conventional ESTJ model, appears as an additive term. But, if transaction costs are not accounted for while estimating the model, as is the case with many of the abovementioned studies, this variable enters the error term. Transaction cost contains various components like transportation cost, variable costs (insurance, hedging, contracting etc.), duties levied and other unmeasurable costs (search costs, risk premia etc.). By their very nature, components like insurance costs and ad-valorem duties depend on price. When these components enter the error term, it is a violation of the basic assumption of the general linear model (that the explanatory variables and error term are uncorrelated). This will yield inconsistent estimates. Also, given the emphasis on tariff reduction in international negotiations, ad-valorem duties may be non-stationary violating yet another assumption of the classical model. Also, the case of discontinuous trade poses a serious problem. The standard ESTJ model envisages multiple price equilibria. However, an overwhelming majority of the empirical models in the literature impose an equality constraint and test for a single equilibrium. Analyses of markets marked by discontinuous trade are flawed unless multiple equilibria are accounted for.

Some noteworthy extensions of cointegration methods have evolved to address the issue of non-tradability regions in the ESTJ models by employing threshold autoregression and threshold cointegration methods (Obstfeld and Taylor, 1997; Prakash and Taylor, 1997, Goodwin and Grennes, 1998; Dercon, 1995; Goodwin and Piggott, 2001). The non-linearity in price adjustment, because of transaction cost, prompted research in two different directions. Threshold autoregression or TAR models, accounting for threshold effects, and switching regime models (Sexton *et al.,* 1991). The switching regimes models was further extended by Baulch (1997) to include transaction cost data to formulate what is popularly called the parity bounds model (PBM). The findings of these models suggest that a careful accounting of the unobservable threshold effects due to transaction costs reduces the likelihood of incorrectly rejecting the market efficiency hypothesis. When price data alone are available, this seems like a sensible way to proceed.

Methods using mixture distribution models have attempted to address the issues of discontinuous trade and non-stationary transactions cost data. These models interpret prices as being drawn from three distinct regimes—first, where price differentials are less than the transaction costs; second, where the two are equal; and the third, where the price differential exceeds transaction cost (Spiller and Wood, 1988; Sexton *et al.*, 1991; Baulch, 1997; Barrett and Li, 2002; Barrett *et al.*, 2000). While Spiller and Wood (1988) and Sexton *et al.* (1991) use variants of switching regimes model that relies on price data, Baulch (1997) developed a PBM model by integrating price and transactions cost data. Barrett *et al.* (1999) extended PBM further to make use of trade data. Like the threshold autoregression (TAR) and threshold cointegration (TCI) models, the PBM and BLM models also find greater support for the hypothesis of market efficiency if proper account is taken of transaction cost and trade flow patterns. A good review of some of the recent developments in market integration can be found in Campenhaut (2007).

Threshold Autoregression Model (TAR)

Threshold autoregression models are similar to regime switching models with one difference in that the regime switch is triggered when a forcing variable (or variables) crosses a threshold. In the models that include autoregressive structures, a sequence of observations at any point in time may involve various combinations of alternative regimes, making the underlying structure of the model intrinsically non-linear.

Let $m_t = p_t - p_{rt}$ where p_t is the price in the present market at t and p_{rt} is the price in the reference market r at t.

$$\Delta m_t = \rho m_{t-1} + \varepsilon_t \text{ where } \Delta m_t = m_t - m_{t-1}, \varepsilon_t \sim N(0,\sigma^2) \text{ is the residual.}$$

One of the simplest TAR models (the SETAR model) is given by:

$$\Delta m_t = \rho_{out} m_{t-1} + \varepsilon_t \quad ; \quad m_{t-1} > \theta$$

$$\Delta m_t = \rho_{in} m_{t-1} + \varepsilon_t \quad ; \quad -\theta \le m_{t-1} \le \theta$$

$$\Delta m_t = \rho_{out} m_{t-1} + \varepsilon_t \quad ; \quad m_{t-1} < -\theta$$

where θ is the transaction cost.

Two adjustment parameters—one inside the band (ρ_{in}) and one outside the band (ρ_{out}) are estimated. Inside the band, that is, when the price difference is less than or equal to the transaction cost, there is no trade. In this case, the best guess of m_t is m_{t-1} (naïve expectation). Using this property, inside the band it is $\Delta m_t = \varepsilon_t$. The model then reduces to:

$$\Delta m_t = \rho_{out} m_{t-1} + \varepsilon_t \quad ; \quad m_{t-1} > \theta$$

$$\Delta m_t = \varepsilon_t \quad ; \quad -\theta \le m_{t-1} \le \theta$$

$$\Delta m_t = \rho_{out} m_{t-1} + \varepsilon_t \quad ; \quad m_{t-1} < -\theta$$

A good application of TAR model to test for market integration can be found in Goodwin and Piggott (2001). Tsay (1989) developed techniques of using autoregressive models for testing threshold effects and modelling TAR processes. An improvement of self-exciting threshold autoregressive (SETAR) model can be made by making both the transaction cost (θ) and the adjustment parameter (ρ_{out}) functions of time to account for the changing (decreasing) transaction costs over time due to improvements in transportation, communication etc. (Campenhout, 2007). A superb exposition of SETAR models can be found in Sephton (2003).

Balke and Fomby (1997) recognised the correspondence between error correction models representing cointegrating relationships and the autoregressive models of the error correction term. For instance, consider a simple equilibrium relationship of the form $y_t = \alpha + \beta x_t + e_t$. The extent of cointegration between y_t and x_t depends upon the autoregressive root ρ in $e_t = \rho e_{t-1} + \upsilon_t$. As ρ approaches 1, deviations from equilibrium become non-stationary and thus the two series become non-cointegrated. Using the corresponding error correction representation, one can get $\Delta e_t = \lambda e_{t-1} + \eta_t$ where $\lambda = 0$ when $\rho = 1$.

Using this simple property, Balke and Fomby (1997) extended the TAR models to a cointegration framework. They also showed that the standard methods of testing for unit roots and cointegration also work reasonably well even in the presence of threshold effects. The general model of threshold cointegration developed by Balke and Fomby is as follows. Let the standard cointegrating relationship be given by:

$y_{1t} - \beta_1 y_{2t} - ...... - \beta_k y_{k+1,t} = \upsilon_t$ where $\upsilon_t = \rho \upsilon_{t-1} + e_t$. As already discussed above, the cointegrating relationship among y_{it} variables depends upon the nature of autoregressive pattern of υ_t. As ρ approaches 1, deviations from equilibrium become non-stationary and the y_{it} variables are not cointegrated. Balke and Fomby extended this simple framework to cases where υ_t follows a threshold autoregression.

$$\rho = \rho^{(1)} \text{ if } \quad |\upsilon_{t-d}| \le c$$

$$= \rho^{(2)} \text{ if } \quad |\upsilon_{t-d}| > c$$ where 'd' is the delay parameter and c is the threshold which delineates alternate regimes.

$\rho^{(1)} = 1$ implies that the long-run relationship among y_{it} variables is characterised by a random walk when deviations from equilibrium are small ($\le c$).

We assume d=1 here. The corresponding error correction framework is as follows:

$$\Delta y_t = \sum_{i=1}^{p} \gamma_i^{(1)} \Delta y_{t-i} + \theta^{(1)} \upsilon_{t-1} + \varepsilon_t^{(1)} \text{ if } |\upsilon_{t-1}| \le c$$

$$= \sum_{i=1}^{q} \gamma_i^{(2)} \Delta y_{t-i} + \theta^{(2)} \upsilon_{t-1} + \varepsilon_t^{(2)} \text{ if } |\upsilon_{t-1}| > c$$

This framework can be easily generalised to multiple thresholds, implying multiple regimes allowing for asymmetric adjustment. In general, when there are k thresholds there are k+1 distinct regimes.

Tsay (1989) developed a general non-parametric test for testing non-linearity implied by the thresholds. Let the standard AR model be of the form $\upsilon_t = \alpha + \gamma \upsilon_{t-1} + \varepsilon_t$. Each combination of υ_t and υ_{t-1} is treated as a data case. The data cases are then ordered according to the variable of interest i.e., υ_{t-1}. The model is estimated initially for a sample of k observations (k<n). Then recursive residuals are obtained by sequentially estimating the model by adding one observation at a time. A test of non-linearity is then given by the F-statistic obtained from regressing the recursive residuals on the explanatory variables, υ_{t-1}. Obstfeld and Taylor (1997) suggested that the test be carried out increasing and decreasing orderings of the data cases. Tsay's test is also useful in determining the delay parameter d. The test is run for alternative delay parameters and the d that gives the largest F-statistic is chosen as optimal.

A good application of TAR approach to test agricultural market integration can be found in Goodwin and Piggott (2001) model. In GP model, first the standard DF tests for unit roots and the Johansen cointegration are carried out to assess the level of integration and the presence of cointegration. Then the Engle-Granger method is used to estimate the cointegrating relationship among the variables. Lagged residuals from these regressions define the error correction terms. Following this, the price differentials and the error correction terms from the previous regressions are used to carry out the Tsay's test for the presence of threshold behaviour. Once threshold behaviour is identified, a two-dimensional grid search is carried out to define the thresholds c1 and c2. The cointegration and error correction model is then re-estimated conditional on the threshold parameters. Finally, a test to assess parameter differences across regimes (Hansen, 1997), similar to the standard Chow test, is carried out.

Parity Bounds Model (PBM)

The original PBM identifies three exhaustive regimes based on the price differences between two markets. Either the price difference is equal to the transaction cost (regime 1) or greater than the transaction cost (regime 2) or lesser than the transaction cost (regime 3). In the switching regimes model, from which PBM is derived, regime 1 is modelled as a constant (transactions cost) plus a normally distributed error term. For regime 2, an additional error term is added while for regime 3 an additional error term is subtracted. This additional error term is assumed to follow a truncated half-normal distribution, truncated from below at zero. After formulating the corresponding density functions for each regime, probabilities are assigned to each regime and likelihood functions are specified. Maximising the logarithms of the likelihood function yields estimates of the probabilities of being in one of the three regimes, transaction cost and the standard errors of the two error terms.

Limitations of TAR and PBM Models

All these methodological innovations still suffer from some limitations. The TAR and TCI models continue to rely on strong assumptions about stationarity of transaction costs or continuity of trade flows or both. On the other hand, PBM and BLM models offer only static comparisons. They assume serial independence of price and transactions cost data and therefore do not allow analysis of the dynamics of short-run deviation from long-run equilibrium. Also, these switching regimes models are sensitive to underlying distributional assumptions (Fackler and Goodwin, 2001).

PBM models in particular have some specific limitations. Suppose the price difference is larger than the transaction cost (regime 2). Profitable arbitrage opportunities exist in this regime and thus the probability of observing large deviations from transaction cost is lower than that of observing small deviations. Therefore in this case, the assumption of half-normal distribution is justified. But in the case of regime 3, when the transaction cost is higher than the price difference and there is no trade, there is no reason to expect a higher probability for smaller deviations as compared to larger deviations. It is pertinent here to note that the original switching regimes model of Sexton *et al.* (1991) was developed for regions for which there was trade. The interpretation was that there was too much trade (glut) leading to the price difference falling below the transactions cost. In such cases where trade occurs, the assumption of half-normal distribution can be justified. But in a number of subsequent studies applying switching regimes model and PBM to cases where trade does not occur and price difference is lower than the transaction cost, the assumption of half-normal distribution appears questionable. Even in cases where the markets are connected by trade, the assumption that the adjustment in regimes 2 and 3 is the same appears questionable in view of the following leverage effect (Deaton and Laroque, 1992). Suppose the traders possess the ability to hold stocks and the price difference falls below the transaction cost (regime 3), traders may prefer to hold stocks than sell. In this case the probability of larger deviations can be expected to be higher than that of smaller deviations, again making the assumption of half-normal distribution untenable.

The second major limitation of PBM models is that they are static in nature: the PBM indicates the probabilities of being away from the parity bounds but does not tell us anything about the persistence of these deviations. Using such a static model, it is possible to observe a high frequency of inefficient arbitrage and reach a conclusion that markets are poorly integrated.

Use of price data alone may lead to rejection of the hypothesis of market efficiency due to the presence of transaction costs. Therefore, a finding in support of the null hypothesis with a parsimonious model using only price data may strengthen the case in favour of market efficiency (Barrett, 2001).

Market Integration Studies in India

In the literature in the Indian context, most of the studies do not explicitly make the conceptual distinction between integration and competitive equilibrium (*a la* Barrett and Li, 2002).[1] However, the non-inclusion of transaction costs and trade flows data makes it implicitly clear that all these studies are concerned primarily with the market efficiency aspect as captured by the competitive price equilibrium.

Many early empirical studies of market integration, particularly in the developing countries like India, used simple bivariate correlation coefficient between prices as a measure of market integration. Perhaps the first attempt to use this approach is that of Mahendru (1937). Mahendru analysed wheat prices in Punjab. His correlation coefficients ranged from 0.43 to 0.86 for fortnightly prices observed over a period of six months at four important wheat markets in Punjab. Jasdanwalla (1966) used monthly groundnut prices of terminal and local markets to analyse spatial price linkages. His conclusion was that the correlation coefficients among

1. One exception is Sekhar (2008).

terminal markets were stronger as compared to those among local markets and terminal markets. Other important early works using bivariate correlation coefficients in the Indian context are those of Lele (1968) and Cummings (1967). Lele analysed weekly wheat prices for five Punjab market towns and Delhi while Cummings analysed the Indian wheat market. All the above mentioned studies suffer from the limitation that they use raw price series data and as a result the calculated price correlations are biased upwards. The raw price series are more likely to include the influences of common factors such as climate patterns, inflation or any other shocks that affect all the markets. This is likely to conceal the true degree of integration among markets being analysed. Harris (1979) pointed out that the aggregation of individual price series over time may make these systemic effects more pronounced. She points out that a finely sampled data are likely to be more revealing of the actions of the traders.

A methodologically correct way of using correlation coefficients in spatial price analysis is to calculate the correlation coefficient of the residuals (Croxton and Cowden, 1955), after the trend and the seasonal components are purged from the raw price data. Blyn (1973) makes this correction and reworks the correlation coefficients after eliminating trend and seasonal influences using the same eight-year monthly wheat price data for eight Punjab markets and Delhi as used in Cummings (1967). The seasonal component is removed by arranging the data into 12 monthly series for each market. Then a linear trend line is fitted to each of these given-month (monthly) series and residuals are obtained. Then correlation coefficient is calculated among these residuals (detrended data) of each month for these nine markets. His calculations show that the average correlation coefficient is much lower than reported by Cummings.

The lower correlation coefficients can be explained mainly on the basis of transaction costs. An arbitrageur will be interested in buying from locations with low prices and selling in regions of high prices. But this would require, at the time of the study, knowledge of at least 86 important markets in Punjab and 232 markets registerd for produce assembling (Blyn, 1973). This is a formidable requirement indeed. Even assuming that such knowledge is available, a non-local buyer will require reliable local agents and efficient transport infrastructure to make use of the arbitrage opportunities. These transaction costs impede the degree of correlation or comovement among various markets. On the other hand, monopoly procurement from different markets at fixed prices may lead to high correlations. Other basic problems with correlation approach are the heterogeneity of goods, uncertainty and inaccurate and missing price data (Harris, 1979).

With the advances in time series and econometric techniques, recent studies in the Indian context have started using the cointegration methodology (Behura and Pradhan, 1998; Wilson, 2001; Kumar and Sharma, 2003). Behura and Pradhan (1998) use Engle-Granger cointegration to test for long-run equilibrating relationship among fish prices in six markets in the state of Orissa. The limitations of Engle-Granger method in case of more than two variables are well established and therefore not discussed here.

Wilson (2001) analysed the degree of price interdependence among markets using the monthly wholesale prices of wheat, *jowar*, paddy rice, groundnut and rapeseed-mustard. Wilson analysed the degree of integration among different markets both before and after liberalisation (in 1992) using the Johansen cointegration method. The results of the study show increasing degree of market integration for wheat, paddy rice and rapeseed-mustard crops in the post-liberalisation period while for *jowar* and groundnut the results are opposite. The major drawback of Wilson's study is the categorisation of pre- and post-liberalisation period. Although structural reforms were initiated in Indian economy in the early 1990s, Indian agriculture remained virtually insulated until 1994, when edible oil imports were liberalised followed by relaxation of restrictions on exports of rice in 1995. Therefore, Wilson's analysis using 1992 as the break point is open to question. Other critiques of the cointegration methodology like the assumption of stationary transaction costs, which underpins the exclusion of such costs from the analysis, and the non-consideration of discontinuous trade, make the findings of this study at best suggestive.

Kumar and Sharma (2003) also used the Johansen cointegration method to assess the market efficiency at macro level and multiple regression analysis at the micro level to test for integration of paddy markets in Haryana. Their findings suggest that, at the macro level, the four regulated markets in the state are well-integrated. However, their findings at the micro level are quite unsatisfactory due to the very low explanatory power of the models fitted.

Despite the increasing recognition of the role of transaction costs in market integration, even the most recent studies in Indian context have not incorporated these into their analyses (Jha *et al.*, 2005; Jayasuriya, 2008; Sekhar, 2008). The probable reason is the non-availability and non-reliability of transaction costs data.

Jha *et al.* (2005) tested for market integration in 55 wholesale rice markets in India using monthly data from 1970 to 1999. They have used the Gonzalez-Riviera and Helfand (2001) model. Their findings suggest that the market integration in India is incomplete probably because of excessive government interference in rice markets.

Jayasuriya *et al.* (2008) examined the nature of international price transmission into Indian markets before and after the liberalisation in 1994. Their findings suggest that the integration of domestic market with international market improved significantly after the liberalisation. Their methodology is slightly different from most of the earlier studies and some discussion is in order here. They have used panel unit root methods in combination with half life-cycle estimation to assess the nature and degree of market integration in Indian rice markets. The panel unit root methods are expected to exploit the larger (and richer) panel datasets. However, there are some basic limitations to this approach. Firstly, panel unit root tests can not disentangle the speed of convergence of individual time series (markets), which is a very important policy issue in market integration. Secondly, half-life estimation is based on impulse response functions (IRF) and IRFs are not uniquely distinguishable when shocks to the system are correlated, which is precisely the case in spatially separated markets. A shock to one market is expected to be transmitted to other markets as well, depending on the degree of integration. Also, IRFs are not invariant to the ordering of the variables.

To address these limitations, Sekhar (2008) used a different methodology. The predominant approach in the literature on spatial market integration is of cointegration and error correction. The basic idea of cointegration approach is that there is a common stochastic trend among variables (here prices) and purging that common trend, the long-run equilibrium relationship (among prices at different locations) can be identified. In this approach, each price is expressed as a function of all the prices. However in this approach, the specific markets contributing to the price movements cannot be identified.

Therefore, Sekhar (2008) used a common factor representation (Stock and Watson, 1988) of the cointegrated series, called Gonzalo-Granger (G-G) model. In this approach, linear combination(s) of a subset(s) of prices, acting as a common factor(s), underlie the entire constellation of prices. Unlike the cointegration approach, in this method the common stochastic factor is not eliminated. The empirical estimation of the model is also relatively simple and does not call for new methods. One can estimate the standard vector error correction (VEC) model and then use those estimates to: i) determine the number of common factors, ii) identify the specific markets influencing price formation (Gonzalez and Helfand, 2001; Gonzalo-Granger, 1995). A market is said to be integrated if there exists a single common factor (single linear combination of a subset of prices). The assumptions made in the model are: i) the model assumes centrality of a subset of markets and stationary transaction/transportation costs, ii) the common factors are linear combinations of observables i.e., prices and iii) the common factors are not affected by cyclical (stationary) movements in prices

The following specific steps are involved in assessing the extent and degree of market integration in this methodology—a) identifying a common integrating factor based on the Gonzalo and Granger (1995) methodology. This methodology is particularly attractive because it allows identification of the location(s) that contribute to the long-run behaviour of the prices through the common integrating factor, which is based on observable variables. b) measuring the degree of market integration, defined as the reaction time to remove the disequilibrium or shock, by means of 'persistence profiles' (Pesaran and Shin, 1996). A persistence profile characterises the response of a cointegrating relation to a system-wide, rather than, to an individual shock. It measures the reaction time of each of the long-run equilibrium relations to absorb a system-wide shock.

Using this methodology, Sekhar tested for market integration for a number of commodities in India. His findings are similar to that of Jha *et al.* (2005) in case of rice market, that the integration at the national level is not satisfactory although at the state and regional levels it is complete. He finds that the markets of the commodities facing lesser trade restrictions domestically, like edible oils and pulses, are better integrated as compared to rice, which faces the maximum interstate restrictions.

Conclusions

In conclusion, conceptual distinction between market integration and market efficiency is still elusive in the literature. There is a need for researchers to make this distinction explicit in their studies. Also, the need to develop and make available accurate transactions cost data cannot be overemphasised for meaningful research on market integration. In the Indian context, the availability and quality of even basic price data across markets leaves a lot to be desired. Reliable time series data on prices and transaction costs across various important markets in the country can go a long way in improving the quality of market integration studies. Such a dataset can allow application of the recent advances in estimation techniques like the TAR and PBM models to study the degree of market integration in India.

References

Alexander, C. and J. Wyeth (1994). "Cointegration and Market Integration: An Application to the Indonesian Rice Market", *Journal of Development Studies* 30 (2): 303–328.

Ardeni, P.G. (1989). "Does the Law of One Price Really hold for Commodity Prices?", *American Journal of Agricultural Economics* 71(3): 661–669.

Badiane, O. and G.E. Shively (1998). "Spatial Integration, Transport Costs and the Response of Local Prices to Policy Changes in Ghana", *Journal of Development Economics* 56: 411–431.

Balke, N.S. and T.B. Fomby (1997). "Threshold Cointegration", *International Economic Review* 38(3): 627–645.

Barrett, C. (1996). "Market Analysis Methods: Are Our Enriched Toolkits Well Suited to Enlivened Markets?", *American Journal of Agricultural Economics* 78: 825–9.

———. (2001). "Measuring Integration and Efficiency in International Agricultural Markets", *Review of Agricultural Economics* 23: 19–32.

Barrett, C. and J. Li (2002). "Distinguishing between Equilibrium and Integration in Spatial Price Analysis", *American Journal of Agricultural Economics* 84: 292–307.

Barrett, C.B., J.R. Li and D. Bailey (1999). "Factor and Product Market Tradability and Equilibrium in Pacific Rim Pork Industries", *Journal of Agricultural and Resource Economics* 25: 68-87. July.

Baulch, B. (1997). "Transfer Costs, Spatial Arbitrage, and Testing for Food Market Integration", *American Journal of Agricultural Economics* 79: 477–87.

Behura, Debdutt and Durga Charan Pradhan (1998). "Cointgration and Market Integration: An Application to the Marine Fish Markets in Orissa", *Indian Journal of Agricultural Economics* 53(3): 344-350. July-September.

Blyn, G. (1973). "Price Series Correlation as a Measure of Market Integration", *Indian Journal of Agricultural Economics* 28: 56-59.

Campenhout, Bjorn Van (2007). "Modelling Trends in Food Market Integration: Method and an Application to Tanzanian Maize Markets", *Food Policy* 32: 112-117.

Croxton, Frederick E. and Dudley J. Cowden (1955). *Applied General Statistics*, Second Edition. New York: Prentice Hall.

Cummings, Ralph W., Jr (1967). *Pricing Efficiency in the Indian Wheat Market*. New Delhi: Impex.

Deaton, A., G. Laroque (1992). "On the Behaviour of Commodity Prices", *Review of Economic Studies* 59(1): 1–23.

Dercon, S. (1995). "On Market Integration and Liberalisation: Method and Application to Ethiopia", *Journal of Development Studies* 32. October.

Enke, S. (1951). "Equilibrium Among Spatially Separated Markets: Solution by Electrical Analogue", *Econometrica* 19: 40–7.

Fackler, P.L. and B.K. Goodwin (2001). "Spatial Price Analysis", in B.L. Gardner and G.C. Rausser (eds.), in *Handbook of Agricultural Economics, Marketing, Distribution and Consumption*, Vol. 1B. Amsterdam: Elsevier/North-Holland. pp.971–1024.

Fafchamps, M. and S. Gavian (1996). "The Spatial Integration of Livestock Markets in Niger", *Journal of African Economies* 5(3): 366–405.

Goletti, F. (1994). "The Changing Public Role in a Rice Economy Approaching Self-Sufficiency: The Case of Bangladesh", *Research Report* No.98. Washington, DC.: International Food Policy Research Institute.

Goodwin, B.K. and T.J. Grennes (1998). "Tsarist Russia and the World Wheat Market", *Explorations in Economic History* 35: 405-30.

Goodwin, B.K. and N.E. Piggott (2001). "Spatial Market Integration in the Presence of Threshold Effects", *Journal of Agricultural Economics* 83(2): 302–317.

Goodwin, B.K. and T.C. Schroeder (1991). "Cointegration Tests and Spatial Price Linkages in Regional Cattle Markets", *American Journal of Agricultural Economics* 73(2): 452–464.

Gonzalez-Riviera, G. and S.M. Helfand (2001). "The Extent, Pattern and Degree of Market Integration: A Multivariate Approach for the Brazilian Rice Market", *American Journal of Agricultural Economics* 83(3): 576-592.

Gonzalo, J. and C.W.J. Granger (1995). "Estimation of Common Long Memory Components in Cointegrated Systems", *Journal of Business & Economic Statistics* 13(1): 27-35. January.

Gupta, S. and R. Mueller (1982). "Analysing the Pricing Efficiency in Spatial Markets; Concept and Application", *European Review of Agricultural Economics* 9: 24-40.

Hansen, B.E. (1997). "Inference in TAR Models", *Studies in Nonlinear Dynamics Econometrics* 2, April (online).

Harris, B. (1979). "There is Method in My Madness: Or is it Vice Versa?: Measuring Agricultural Market Performance", *Food Research Institute Studies* 17: 197-218.

Hurd, J. (1975). "Railways and the Expansion of Markets in India, 1861-1921", *Explorations in Economic Hisory* 12: 263-88.

Jasdanwalla, Zaibun Y. (1966). *Marketing Efficiency in Indian Agriculture*. Allied Publishers. pp.106-108

Jayasuriya, Sisira, Jae K. Him and Parmod Kumar (2008). "Opening to World Markets: A Study of Indian Rice Market Integration". Paper presented at the NCAER-ACIAR Conference. New Delhi. 6 June.

Jha, R., K. Murthy, H. Nagarajan and A. Seth (2003). "Market Integration in Indian Agriculture", *Economic Systems* 21(3): 217–34.

Jha, R., K.V. Bhanumurthy and Anurag Sharma (2005). "Market Integration in Wholesale Rice Markets in India", *ASARC Working Paper* 2005/03.

Jones, W.O (1968). "The Structure of Staple Food Marketing in Nigeria as Revealed by Price Analysis", *Food Research Institute Studies* 8: 95-124.

Kumar, Parmod and R.K. Sharma (2003). "Spatial Price Integration and Pricing Efficiency at the Farm Level: A Study of Paddy in Haryana", *Indian Journal of Agricultural Economics* 58(2): 201-217. April-June.

Lele, U. (1967). "Market Integration: A Study of Sorghum Prices in Western India", *Journal of Farm Economics* 49: 149-59. February.

———. (1968). *Working of Grain Markets in Selected States of India: 1955-56 to 1964-65*. Department of Agricultural Economics. New York: Cornell University, Ithaca. December.

Mahendru, I.D. (1937). "Some Factors Affecting the Price of Wheat in Punjab", *Board of Economic Enquiry Publication* No. 49. Punjab Government.

Obstfeld, M. and A.M. Taylor (1997). "Nonlinear Aspects of Goods-Market Arbitrage and Adjustment; Heckscher's Commodity Points Revisited", *Journal of the Japanese and International Economies* 11(4): 441–479.

Palaskas, T.B. and B. Harriss-White (1993). "Testing Market Integration: New Approaches with case material from the West Bengal Food Economy", *Journal of Development Studies* 30 (1): 1–57.

Pesaran, M.H. and Y. Shin (1996). "Cointegration and Speed of Convergence to Equilibrium", *Journal of Econometrics* 71: 117-43.

Prakash, Gauri and Alan M. Taylor (1997). "Measuring Market Integration: A Model of Arbitrage with an Econometric Application to the Gold Standard, 1879-1913", *NBER Working Paper* No. 6073. June.

Rashid, S. (2004). "Spatial Integration of Maize Markets in Post-Liberalised Uganda", *Journal of African Economies* 13 (1): 102–133.

Ravallion, M. (1986). "Testing Market Integration", *American Journal of Agricultural Economics* 68: 102–9.

Samuelson, P. (1952). "Spatial Price Equilibrium and Linear Programming", *American Economic Review* 42: 283–303.

Sekhar, C.S.C (2008). "Agricultural Market Integration in India: An Analysis of the Select Commodities", in *How are the Poor Affected by International Trade in India: An Empirical Approach*. UNCTAD-MOIC-DFID Report on "Strategies and Preparedness for Trade and Globalisation in India".

Sephton, P.S. (2003). "Spatial Market Arbitrage and Threshold Cointegration", *American Journal of Agricultural Economics* 85(4): 1041–1046.

Sexton, R.J., C.L. Kling and H.F. Carman (1991). "Market Integration, Efficiency of Arbitrage and Imperfect Competition: Methodology and Application to US Celery", *American Journal of Agricultural Economics* 73 (3): 568–580.

Spiller, P. and R.O. Wood (1988). "The Estimation of Transaction Costs in Arbitrage Models", *Journal of Econometrics* 39: 309-326.

Stock, J.H. and M.W. Watson (1988). "Testing for Common Trends", *Journal of American Statistical Association* 83: 1097-1107

Takayama, T. and G. Judge (1971). *Spatial and Temporal Price Allocation Models*. Amsterdam: North-Holland.

Taylor, A.M. (2001). "Potential Pitfalls for the Purchasing-power-Parity Puzzle? Sampling and Specication Biases in Mean-reversion Tests of the Law of One Price", *Econometrica* 69 (2): 473–498.

Timmer, C.P. (1989). "Food Price Policy: The Rationale for Government Intervention", *Food Policy* 14(1): 17-42.

Tsay, R.S. (1989). "Testing and Modeling Threshold Autoregressive Processes", *Journal of American Statistical Association* 84: 231-40. March.

Wilson, Edgar J. (2001). "Testing Agricultural Market Integration: Further Conceptual and Empirical Considerations Using Indian Wholesale Prices", in S.S. Acharya and D.P. Chaudhari (ed.), *Indian Agricultural Policy at the Cross-Roads*. New Delhi: Rawat Publishers.

Yavapolkul, Navin, Munisamy Gopinath and Ashok Gulati (2004). "Post Uruguay Round Price Linkages Between Developed and Developing Countries: The Case of Rice and Wheat Markets", *MTID Discussion Paper* No. 76, Washington, D.C.: International Food Policy Research Institute.

6 Vertical Coordination in Agribusiness in India

Making Contract Farming Work for Small Producers

Sukhpal Singh

Linking farmers with markets is a major thrust of agribusiness policy across the developing world including India under the new economic environment of liberalisation and globalisation. Contract farming (CF) is central to this thrust as it brings corporate agribusiness entities in direct contact with the primary producers in the form of a partnership for betterment of both the parties. Due to the efficiency (coordination and quality control in a vertical system) and equity (smallholder inclusion) benefits of this hybrid system, it has been promoted aggressively in the developing world by various agencies (Glover, 1987). CF can be defined as a system for the production and supply of agricultural and horticultural produce by farmers/primary producers under advance contracts, the essence of such arrangements being a commitment to provide an agricultural commodity of a type, at a specified time, place of delivery/pick up, price, and in specified quantity to a known buyer. It basically involves five things —pre-agreed price, quality, quantity or acreage (minimum/maximum), place of delivery and time (Singh, 2002).

CF is known by different variants like centralised model which is company-farmer arrangement, outgrower scheme which is run by government/public sector/joint venture, nucleus-outgrower scheme involving both captive farming and CF by the contracting agency, multi-partite arrangement involving many types of agencies, intermediary model where middlemen are involved between the company and the farmer, and satellite farming referring to any of the above models (Eaton and Shepherd, 2001; Singh, 2008c). In fact, CF varies depending on the nature and type of contracting agency, technology, nature of crop/produce, and the local and the national context.

The contracts could be of three types: (i) procurement contracts under which only sale and purchase conditions are specified; (ii) partial contracts wherein only some of the inputs are supplied by the contracting firm and the produce is bought at pre-agreed prices and (iii) total contracts under which the contracting firm supplies and manages all the inputs on the farm and the farmer becomes just a supplier of land and labour. The relevance and importance of each type varies from product to product and over time and these types are not mutually exclusive (Hill and Ingersent, 1987; Key and Runsten, 1999). Whereas the first type is generally referred to as marketing contracts, the other two are types of production contracts (Scott, 1984; Welsh, 1997). But, there is a systematic link between product and factor markets under the contract arrangement as contracts require definite quality of produce and, therefore, specific inputs (Scott, 1984). Also, different types of production contracts allocate production and market risks between the producer and the processor in different ways. The price of the contracted produce can be growers' fixed price, residual (profit/loss) sharing by sponsor and grower, open market price based price, spot market price, consignment based, two-part split price, tournament price (fixed plus variable based on relative performance), base price plus quality based incentive price, or administered price.

This paper examines the rationale and relevance of CF in India and examines its inclusiveness and the ways and means of making it accessible to small producers who are the predominant majority in India. Section II provides the rationale for CF followed by its small producer relevance, interface and the issues involved in Section III. Section IV examines the practice of CF in India so far as its inclusiveness or absence of it is concerned. Section V focusses on making contract farming relevant and accessible for small producers from various perspectives like policy and institutional. Section VI concludes the paper with main issues and strategies to make CF accessible to and affordable for small producers.

Rationale for CF

A farmer may prefer a contract as it gives access to additional sources of capital (credit), and a more certain market and price by shifting a part of the risk of adverse price movement to the buyer (Wolf *et al.*, 2001; Simmons *et al.*, 2005). Farmers also get an access to new technology and inputs through contracts which otherwise may be outside their reach (Eaton and Shepherd, 2001). In fact, CF is also seen as an institutional innovation to help farmers and agricultural development (Weatherspoon *et al.*, 2001).

For agribusinesses, CF is an alternative to corporate farming which may be costly, risky, non-available as an option, difficult to manage and not viable (Simmons *et al.*, 2005) in competitive markets. Contracts also help improve product quality by directly introducing incentives and penalties as there are problems of adverse selection and moral hazard in any contractual arrangement resulting in underinvestment or shirking by any of the parties (Wolf *et al.*, 2001; Simmons *et al.*, 2005).

CF is also suggested as the only way to make small scale farming competitive as the services provided by contracting agencies can not be provided by any other agencies (Eaton and Shepherd, 2001). Contract farming lowers transaction costs for the farmers as well as processors or marketers as many of the transactions are internalised (Wolf *et al.*, 2001). Transaction costs are defined as costs of running the contract system and include *ex-ante* costs of drafting, negotiating and safeguarding the agreement and *ex-post* cost of maladaptation and of dealing with it, set up and running costs of governance structures and bonding costs of securing commitments. These costs arise from bounded rationality, opportunism and asset specificity (Simmons *et al.*, 2005; Sartorius and Kirsten, 2005).

At more macroeconomic level, contracting can help to remove market imperfections in produce, capital (credit), land, labour, information and insurance markets; facilitate better coordination of local production activities which often involve initial investment in processing, extension etc., and can help in reducing transaction costs (Grosh, 1994; Key and Runsten, 1999). It has also been used in many situations as a policy step by the state to bring about crop diversification for improving farm incomes and employment (Benziger, 1996; Singh, 2000). CF is also seen as a way to reduce costs of cultivation as it can provide access to better inputs and more efficient production methods. The increasing cost of cultivation was the reason for the emergence of CF in Japan and Spain in the 1950s

(Asano-Tamanoi, 1988) and in the Indian Punjab in the early 1990s (Singh, 2000).

From an institutional economics perspective, the logic for CF could also come from the creation of positive externalities like employment, market development or infrastructure, if agribusiness firms create them better than the open market or the state (Key and Runsten, 1999). In other words, CF can help people other than those who have direct stakes and pay for it. Therefore, CF figures as an institutional arrangement for agricultural development in the fields of inputs, product exchange and product upgrading, the last referring to research and innovations (Christensen, 1992).

CF and Small Producers

Given the predominance of small producers in developing economies, small producer participation in CF has been an important issue as small producer inclusion is rather an exception. The aspects of contracting which contribute to CF excluding small producers are: enforcement of contracts, high transaction costs, quality standards, business attitudes and ethics like non/delayed/reduced payment, high rate of product rejection and weak bargaining power of the small growers (Kirsten and Sartorius, 2002).

The organisers of CF also find it costly to work with small producers due to their scattered location and smaller volumes though there are also many advantages of working with them like easy availability of family labour, better commitment levels, helping spread risk of default, giving corporate agency a social face, and lower cost of production due to lower mechanisation of smaller farms (Boselie *et al.*, 2003).

Contracting agencies impose eligibility conditions for contract production like certain land size, no non-contract crop production in the neighbouring fields and certification, especially in seed production (Simmons *et al.*, 2005). The eligibility criteria for participation in CF projects/schemes like irrigated land, suitable land, land near main road, literacy level of the farmer are themselves discriminatory in terms of who can be a contract grower. In fact, in CF everywhere, private agribusiness firms have less interest and ability to deal with small scale farmers on an individual basis (Hazell, 2005). Thus, contracting agencies especially private, tend to prefer large farmers for CF because of their capacity to produce better quality crops due to the efficient and business-oriented farming methods and, large volumes of produce which reduces the cost of collection for the firm and has better traceability due to absence of pooling, their capacity to bear risk in case of crop failure, and various services provided by these large producers like transport, storage, etc., all of which leads to lower cost procurement for the firm (Key and Runsten, 1999; Simmons *et al.*, 2005).

On the other hand, small farmers are picked up by firms for contracts only when the area is dominated by them, there is government directive to do so or they are found to be low cost producers in certain areas and crops (CDC, 1989). Further, firms may work with small farmers to make use of the state support (financial and technical) to these producers under various development programmes (Glover and Kusterer, 1990) and to benefit from lower cost production on these farms as these farmers have access to cheaper family labour, and being residual claimants of their labour, work more conscientiously than hired labour (Key and Runsten, 1999). In fact, some of the agencies even use large growers, rural elite and local small processors as subcontractors to procure from the small growers for the company (Kirk, 1987). The seed companies in India use small companies as subcontractors to procure seeds produced under contracts (Shiva and Crompton, 1998). In Canada, small tomato growers were preferred as the crop required hand-picking which only small farmers do as the large ones do mechanical harvesting, especially when weather limits the use of mechanical harvesters (Winson, 1990). Similarly, in India, gherkin CF is carried out by small and marginal farmers as the crop requires plenty of labour inputs which these farming families can provide from within (Dev and Rao, 2005). Also, working with many small farmers in the case of small processors gives the required flexibility in procurement schedule helping to extend the processing season and use the equipment efficiently; and helps spread risk of supply failure as compared to working with a few large farmers.

Inclusiveness of CF in India

Across crops and regions, most of the firms worked mostly with large and medium farmers (Singh, 2002; Haque, 2003; Dev and Rao, 2005;

Singh and Asokan, 2005; Singh, 2005a; Kumar, 2006) with the exception of firms in Karnataka, Tamil Nadu and Andhra Pradesh which worked with small and marginal farmers due to the nature of the crops (cucumber/ gherkin and broiler chicken) (Dev and Rao, 2005). Surprisingly, even a state sponsored cooperative (Markfed) in Punjab did not entertain farmers who could not spare at least three acres for its *basmati* paddy CF programme (*The Tribune*, Chandigarh, June 8, 2006). In mint CF in Punjab, contract farmers had owned holdings ranging from 5 to 30 acres with the average being 17.18 acres and leased land being of the order of 39.63 acres on an average, making for the average operational holding size of 56.81 acres (Asokan and Singh, 2006) when the average size of holding in the state is just 9 acres. This bias in favour of large/medium farmers is perpetuating the practice of reverse tenancy in regions like Punjab where these farmers lease in land from marginal and small farmers for contract production (Singh, 2002; Haque, 2003).

A recent study of CF operations of Frito-Lay (PepsiCo) and McCain in Punjab and Gujarat (Singh, 2008a) revealed that, by and large, both of the firms studied worked with large or very large growers by Indian standards. Among the sample farmers, only one contract grower with McCain had operational landholdings of less than five acres (although McCain did buy from some small growers without contracts) with average landholding being 19 acres compared with only 5 and 9 acres for farm gate and APMC sellers respectively. Contract growers for Frito-Lay in Punjab had average operational holdings of 63 acres, with only 22 acres owned and the rest leased. None of the sample contract grower with Frito-Lay had less than 10 acres of land, in spite of the fact that the average size of holdings in the state is 9 acres and 70 per cent have holdings below 10 acres each. A recent Punjab based study of contract farming showed that the average size of the operational holding of contract growers was more than one and a half times that of the non-contract growers. It found 'no marginal farmer in the size group of below one hectare [...] operating under contract farming. A handful of small farmers in the size group of one to two hectares was operating' (Kumar, 2006: 5369). Another study noted that 'the majority of the acreage registered in the project (CF by PAFC) is held by larger farmers, who tended to receive greater benefits from participation (in contract farming)' (Witsoe, 2006: 16). Even in organic contract farming projects coordinated by private corporate agencies, most of the growers were large or medium landholders (Singh, 2009) (Table 6.1).

Table 6.1

Crop and Agency-wise Average Size of Landholding of Contract Growers

Study and Year	*Contracting Agency and Place*	*Crop under Contract*	*Average Size of Contract Grower Holding (Acres)*	*Average % Area under Contract*
Singh, 2002	Frito-Lay (PepsiCo), Punjab	Potato	53	8
-do-	Nijjer Agro, Punjab	Tomato	22	23
-do-	HLL, Punjab	Tomato	78	33
-do-	PepsiCo, Punjab	Chillies	90	4.5
Dev and Rao, 2005	AP Govt. and various processors	Oil palm	10	40
-do-	BHC Agro India	Gherkins	7	15
Asokan and Singh, 2006	A.M. Todd, Punjab	Mint	57	-
Kumar, 2006	Many MNCs and local firms in Punjab	Many crops together	37	12
Singh, 2008a	McCain Foods, Gujarat	Potato	19	21
-do-	Frito-Lay (PepsiCo), Punjab	Potato	63	53
-do-	A.M. Todd, Punjab	Mint	40	27
Roopam Singh, 2008	Frito-Lay, (PepsiCo), Punjab	Potato	75	-
Singh, 2009	Agrocel, Haryana	Organic *basmati*	32	-

Source: Various studies.

It is not incidental that most of the CF projects are in the states of Punjab, Haryana, Gujarat, Maharashtra, Karnataka and Tamil Nadu which are agriculturally developed states. On the other hand, vast areas of the country such as Bihar, Jharkhand, Chhattisgarh, Orissa, West Bengal, the entire Northeast India and areas of Uttarakhand, Himachal Pradesh, Kerala and Jammu and Kashmir have been bypassed by CF projects. Does it mean that these areas and farmers would not benefit from commercialisation and vertical coordination of agriculture? These are areas with highest concentration of small and marginal farmers (Gill, 2004). This essentially means that contracting companies do not encourage the participation of those who need to be helped to participate as risk preference and innovativeness require not just attitude but also resources and risk-taking capability to undertake risky crops and ventures (Glover, 1987).

Besides the larger farmer bias of the contracting agencies, the higher cost of production under contract farming also makes it difficult for small holders to sustain their participation especially when there is no credit support provided as part of CF arrangement, besides the higher risk in contract crops which are generally high value crops.

Making CF Work for Small Farmers

There have been warning notes since the 1980s about the widespread applicability of CF as a development tool (Goldsmith, 1985; Glover, 1987). Further, the hybrid nature of CF, and the often conflicting objectives it is supposed to meet, makes it difficult to either design effective public policy interventions or to organise growers (Glover, 1987). Viewed in the light of experience, the potential problems of CF point to the policy steps which are required to reduce the ill-effects of such a system like regulation and monitoring of contracts and to facilitate its smooth functioning in small farmer contexts.

Role of the State and the Development Agencies

The Government of Punjab through Punjab Agro Foodgrains Corporation (PAFC) had been reimbursing for three years extension cost to the CF agencies/facilitators at the rate of Rs 100 per acre. But, doing it irrespective of the size of holding of the contract growers defeated the purpose as it did not ensure that small and marginal farmers who could not afford to pay for extension and needed to be brought into the contract system were included. Similarly, the Ministry of Food Processing Industries (MFPI) has been providing an incentive since the beginning of the Ninth Plan in the form of a reimbursement of five per cent of the value of raw materials procured through CF with farmers with a maximum ceiling of Rs 10 lakh per year for a maximum of three years with the condition that any organisation (private/public/cooperative/NGO/joint venture/assisted) should work with at least 25 farmers under contract for at least three years (MFPI, 1998).

In Thailand, the state not only provided coordination and support of local authorities such as agricultural extension agents, local administration officers and the Bank for Agriculture and Agricultural Cooperatives (BAAC), it also reallocated 250 million Baht deposit in BAAC. The interest compensation for the farmer participants in the programme (3.5% p.a.) was made available to encourage more farmer participation and to reduce production cost. But, later, farmers could obtain only low interest rate (5% p.a.) loan instead of getting compensation for interest charge. The Ministry of Agriculture and Coopeatives (MoAC) through its Department of Agricultural Extension (DoAE) still carries out training in CF for farmers and local officials which includes aspects like guidelines for CF implementation, i.e., types of products suitable for certain kind of contracting arrangement, familiarity with the system of CF, supervision of contract arrangements, special financial assistance to companies undertaking CF, and process of implementation of the CF project in terms of coordination between public and private sectors and choice of relevant area and farmers. It even specifies requirements before a CF project is undertaken, which are: farmer willingness and exposure to the concept of CF, feasible standards of quality of produce, price flexibility and production and marketing plan. Discussing factors in success of CF project, it emphasises on clear understanding of the concept of CF and a felt need for it among project participants, besides physical proximity between producing and processing areas, strict adherence to quality standards and honesty and sincerity of all parties involved. It assigns specific roles and responsibilities to farmers, farmer groups, companies and government agencies in terms of dealing with each other for the project. It finally recommends suitable types of systems of CF for different types of commodities and advocates CF with farmer organisations instead of individual farmers. It even goes on to advise having multiple outlets for produce, both contract and non-contract, for different grades of produce from the same group of farmers (MoAC, 2002). The state intervention helped the farm sector and its farmers only to the extent that it has promoted competition that has been beneficial for growers like in potato in Northern Thailand, and also led to pumping of capital in farm sector through the BAAC loans for contract growers. But, on most other counts, like making the system fair for farmers in terms of nature of contracts, or helping farmer organisations in processing or marketing or even in collective selling to companies, it has not really been effective. The role of the government should be enabling and developmental in nature. That credit support to contract farming projects by the state is crucial has been emphasised earlier as well in other contexts (Goldsmith, 1985).

Regulation of supermarket chains to control or mitigate their market power can be a potential tool to ensure the presence of small growers in value chains as seen in the case of banana trade regime in pre-WTO period in the EU policy, single channel (monopoly) exports by producer bodies in some exporting countries like South Africa, and regulation of domestic import markets in France (Gibbon, 2003). However, regulations do not ultimately change the economic forces under which the supermarkets operate and the changes in procurement systems are driven by these forces. These changes and the basic requirements they impose on growers are conditions which will have to be met if the growers are to be able to tap the powerful market of the supermarkets. Therefore, it is crucial that government and donor agencies help small farmers and entrepreneurs to make the investments in equipment, management, technology, commercial practice and the development of strong and efficient organisations to meet those requirements. There have been such attempts in Brazil and Guatemala (Reardon and Berdegue, 2002). In organic produce chains as well, group certification by public or development agencies can provide a chance for small farmers to take up organic cultivation.

Role of Farmer/Civil Society Organisations

Producers' organisations amplify the political voice of smallholder producers, reduce the costs of marketing of inputs and output, and provide a forum for members to share information, coordinate activities and make collective decisions. Producers' organisations create opportunities for producers to get more involved in value adding activities such as input supply, credit, processing, marketing and distribution. On the other hand, they also lower the transaction costs for the processing/marketing agencies working with growers under contracts. Collective action through cooperatives or associations is important not only to be able to buy and sell at a better price but also to help small farmers adapt to new patterns and much greater levels of competition (Farina, 2002). There is a need to strengthen small farmer organisations and provide them technical assistance to increase productivity for the cost competitive market, provide help in improving quality of produce, and to encourage them to participate more actively in the marketing of their produce in order to capture value added in the supply chain. Finally, the problem of financing the small producers needs to be tackled by finding innovative ways to provide finance (Schwentesium and Gomez, 2002).

There have been cases of success when public or private assistance to the small growers in terms of technical assistance and supply of input credit was made available. In some places in Brazil, small farmers have gone in for collective tanks to meet the scale requirement though the large farmers will have an advantage as they do not face the transaction cost involved in collective use of physical assets. The dairy companies and cooperatives encourage the use of collective tanks, even by financing or facilitating credit for milk producers in some cases (Farina, 2002). Many attempts including offering differentiated contracts under CF schemes to include small farmers failed in Mexico though firms had no option but to contract with small vegetable growers. A few firms succeeded in including small farmers in their CF projects only when they also employed the farmers' family members in their processing units, used local intermediaries to supervise small growers, and/or limited the small grower contracts only to areas close to the highway, besides the other reasons like the small growers being lower cost and more efficient than the firm's own farms, their poor access to alternative outlets for produce and source of credit, and having low labour and land opportunity cost. These measures reduced the information asymmetry between the growers and the firm and the transaction cost of dealing with small growers (Warning *et al.*, 2003). Thus, there is need to promote/encourage farmer groups for CF as was the case in Thailand where besides contract grower groups, the potato growers' cooperative also dealt with a multinational contracting company on behalf of its members. Group CF proves beneficial for both growers and companies (Singh, 2005b) though there may be difficulties in enforcing collective actions due to group heterogeneity, agency resistance to such actions, and making members adhere to group norms in the absence of any legal authority with such collectivities (Glover, 1987).

The groups or farmers' organisations like cooperatives not only lower transaction costs of the firms but also lower input costs for the farmers and give them better bargaining power as was the case of a potato growers' co-operative in North Thailand which acted as a link between the growers and the company (Singh, 2005c). In contract arrangements with small producers in West African countries, the cotton companies started transferring some of the operational or functional responsibilities like distribution of inputs, equipment orders and credit repayment management, to the village associations in the 1970s itself. They provided these associations with management skills for these tasks. The companies relied on traditional village authority structures for organising the associations but limited the associations to one per village to simplify company purchasing, delivery and marketing procedures. This arrangement accounted for a significant part of each cotton company's success (Bingen *et al.*, 2003). But, unfortunately, contracting companies have not been very keen to organise or support cooperatives in India. Producer companies provision of the Companies Act (1956) since 2003 is an opportunity to organise small producers in a cooperative-like but more business-oriented legal entities in India which Tata Chemicals and many NGOs have already taken up recently.

Role of CF Agencies

It is often desirable to plan and provide inputs across a wide range of activities and not just for the raw material needed. Given small land-holdings, it is all the more important that the farmer benefits on a scale that motivates him to do better. This is achieved through a combination of technology, service and networking. Technology involves providing the best inputs and proper harvest and post-harvest conditions. Service ensures that the resources are made available in time and the post-harvest offtake is efficient. Networking assures government support, infrastructure, banking facilities, etc. (Datta, 1996). Partnerships with government or its agencies can help in building relations with the producers. Working with farmer co-operatives for better and more sustainable contract procurement also seems to work. This is being tried out by Marico Industries in Maharashtra in sunflower oilseeds. The company provides working capital, infrastructural facilities, managerial inputs and job work to the cooperatives which helps these oilseed milling cooperatives to become viable.

Conclusions

Though there are concerns about the ability of the small farms and firms to survive in the changing environment of agribusiness, still there are opportunities for them to exploit like in product differentiation with origin of product or organic products and other niche markets. But, the major route has to be through exploitation of other factors like external economies of scale through networking or clustering and such other alliances like CF (Kirsten and Sartorius, 2002). Further, market access for small producers depends on: a) understanding the markets, b) organisation of the firm or operations, c) communication and transport links and d) an appropriate policy environment (Page and Slater, 2003). The experience of CF across the globe suggests that it is not the contract *per se* which is harmful as a system but how it is practised in a given context. If there are enough mechanisms to monitor and use the contract for developmental purposes, it can certainly lead to a betterment of all the parties involved, especially small and marginal farmers.

It is important to remember that there can not be a single blue print or CF model for all situations so far as the role and nature of CF is concerned. Even for individual farmers, it is not contract *per se* but the relationship it represents which is crucial as the divergence between the two may prove crucial in determining the development of CF as an institution (Clapp, 1988; White, 1997). Further, it is the context of the contract which can make a whole lot of difference as there are many actors and factors in the environment which influence the working and outcome of contracts and lead to a culture of contracting which is location and community-specific. The way farmers perceive CF, i.e., define their relationship with the companies, differs in each cultural context (Asano-Tamanoi, 1988; White, 1997). In fact, there is so much diversity in the type of firms, farmers, nature of contracts, crops and socioeconomic environment that it is better to focus on specific situation than the generic institution of CF (Little and Watts, 1994; White, 1997).

Major conditions for successful interlocking between agribusiness firms and small producers include increased competition for procurement instead of monopsony, guaranteed market for farmer produce, effective repayment mechanism, market information for farmers to effectively bargain with companies, large volumes of transactions through groups of farmers, for lowering transaction costs, cooperation among genuine agribusiness firms in the area, and no alternative source of raw material for firms (Kirsten and Sartorius, 2002). Further, for the sustainability of company-farmer partnership schemes, it is important that the company is able to successfully market its products so that farmers do not suffer from lack of market (Baumann, 2000; Haque, 2000). Building of relationships of trust with farmers through company reputation rather than marketing gimmicks is crucial. This requires mutual respect, fair and transparent negotiation process, realistic assessment of benefits, long-term commitment, equitable sharing of risk and sound business plans (Mayers and Vermeulen, 2002). Innovative pricing mechanisms like bonus at the end of the processing cycle, shares in company equity, dividends, producer's fixed price and quality based pricing, which reward performance can help contract performance.

There are a large number of institutional arrangements to coordinate the small producers which should be assessed for their relevance and effectiveness in a given context, though a priori, it seems the cooperative and other similar forms of farmer organisation are more relevant and sustainable, especially the new generation cooperatives (NGCs) which are voluntary, more market-oriented, member responsive, self-governed, and avoid free riding and horizon problems as they have contractual equity based transaction with grower members (Singh, 2008b; Hazell, 2005).

Further, both companies and state should promote group contracts with the intermediation of local NGOs and other organisations and institutions so that contractual relationships are more durable, enforceable, and fair. An insurance component in farming interventions is must to protect the farmer interest and it is noted that some companies are already doing it. But, the most important thing is to ensure market for the farmer produce at better price under these agribusiness projects. Government should also play an enabling role by legal provisions and institutional mechanisms, like helping farmer cooperatives and groups, to facilitate smooth functioning of contract system, and not intervene in CF directly as seen in the case of Punjab where the experiment failed (Kumar, 2006). On the other hand, the success and smooth functioning of the CF system in mint by A.M. Todd in the state (Punjab) with no involvement of the state, due to the nature of the crop, clear terms of the contract, ensured returns to growers by competitive prices and the commitment of the company, corroborates the point that CF is best left to the company and the growers (Singh, 2005). This was also the case in Thailand where the state facilitated it from outside with credit and extension (Singh, 2005c).

Also, it is important to ensure competition in CF so that farmers have choice of options. For example, CF in gherkins in Karnataka was also successful, besides the reason of lack of local market for produce, due to the fact that more than two dozen companies operated in the state. This also helps reduce exploitation of the small growers. Further, since farmers did not put entire land under contract and cultivated only 0.5 acres under gherkin contracts on an average, they were not subject to any major risk of contract failure.

References

Asano-Tamanoi, M. (1988). "Farmers, Industries, and the State: The Culture of Contract Farming in Spain and Japan", *Comparative Studies in Society and History* 30(3): 432-452.

Asokan, S.R. and G. Singh (2006). "Contract Farming of Medicinal Herbs and Organic Crops in India", *CMA Monograph* No. 224. Ahmedabad: CMA, IIM.

Baumann, P. (2000). "Equity and Efficiency in Contract Farming Schemes: The Experience of Agricultural Tree Crops", *ODI Working Paper* No. 139. London: Overseas Development Institute (ODI). October.

Benziger, V. (1996). "Small Fields, Big Money: Two Successful Programs in Helping Small Farmers Make the Transition to High Value-Added Crops", *World Development* 24(11): 1681-1693.

Bingen, J., A. Serrano and J. Howard (2003). "Linking Farmers to Markets: Different Approaches to Human Capital Development", *Food Policy* 28: 405-419.

Boselie, D., S. Henson and D. Weatherspoon (2003). "Supermarket Procurement Practices in Developing Countries: Redefining the Roles of the Public and Private Sectors", *American Journal of Agricultural Economics* 85(5): 1155-1161.

Christensen, S.R. (1992). "Between the Farmer and the State: Towards A Policy Analysis of the Role of Agribusiness in Thai Agriculture", Thailand Development Research Institute (TDRI) Background Report to the 1992 Conference on *Thailand's Economic Structure: Towards Balanced Development?*. Chon Buri, Thailand.

Clapp, R.A.J. (1988). "Representing Reciprocity, Reproducing Domination: Ideology and the Labour Process in Latin American Contract Farming", *The Journal of Peasant Studies* 16(1): 5-39.

CDC (Commonwealth Development Corporation) (1989). *Review of Smallholder Agricultural Programmes, Final Report*, Vols. 1 (analysis, conclusions and proposals) and 2 (case studies). London, CDC, April.

Datta, S.M. (1996). "Linkages with Agriculture: The Experience of the Food Processing Industry", *The Economic Times*. June 26. Ahmedabad.

Dev, S.M. and N.C. Rao (2005). "Food Processing and Contract Farming in Andhra Pradesh: A Small Farmer Perspective", *Economic and Political Weekly* 40(26): 2705-2713.

Eaton, C. and A.W. Shepherd (2001). *Contract Farming: Partnerships for Growth*. Rome: FAO.

Farina, E.M.M.Q. (2002). "Consolidation, Multinationalisation, and Competition in Brazil: Impacts on Horticulture and Dairy Products Systems", *Development Policy Review* 20 (4): 441-457.

Gibbon, P. (2003). "Value Chain Governance, Public Regulation and Entry Barriers in the Global Fresh Fruit and Vegetable Chain into the EU", *Development Policy Review* 21(5-6): 615-625.

Gill, S.S. (2004). "Small Farmers and Markets" (a book review), *Economic and Political Weekly* 39(23): 2356-2358.

Glover, D. (1987). "Increasing the Benefits to Smallholders from Contract Farming: Problems for Farmers' Organisations and Policy Makers", *World Development* 15(4): 441-448.

Glover, D. and K. Kusterer (eds.) (1990). *Small Farmers, Big Business—Contract Farming and Rural Development*. London: Macmillan.

Goldsmith, A. (1985). "The Private Sector and Rural Development: Can Agribusiness Help the Small Farmer?", *World Development* 13(11/12): 1125-1138.

Grosh, B. (1994). "Contract Farming in Africa: an Application of the New Institutional Economics", *Journal of African Economies* 3(2): 231-261.

Haque, T. (2000). "Contractual Arrangements in Land and Labour Markets in Rural India", *Indian Journal of Agricultural Economics* 55(3): 233-252.

———. (2003). "Land Reforms and Agricultural Development: Retrospect and Prospect", in S. Pal *et al.* (eds.), *Institutional Change in Indian Agriculture*. New Delhi: NCAP. 267-284.

Hazell, P.B.R. (2005). "Is There a Future for Small Farms?", in D. Colman and N. Vink (eds.), *Reshaping Agriculture's Contributions to Society*. Proceedings of the 25th International Conference of Agricultural Economists (ICAE). August 16-22. USA: Blackwell. pp.93-101.

Hill, B.E. and K.A. Ingersent (1987). *An Economic Analysis of Agriculture*. London: Heinemann.

Key, N. and D. Runsten (1999). "Contract Farming, Smallholders, and Rural Development in Latin America: The Organisation of Agroprocessing Firms and the Scale of Outgrower Production", *World Development* 27(2): 381-401.

Kirk, C. (1987). "Contracting Out: Plantations, Smallholders and Transnational Enterprise", *IDS Bulletin* 18(2): 45-51, April.

Kirsten, J. and K. Sartorius (2002). "Linking Agribusiness and Small-scale Farmers in Developing Countries: Is There a New Role for Contract Farming?", *Development Southern Africa* 19(4): 503-529.

Kumar, P. (2006). "Contract Farming through Agribusiness Firms and State Corporation: A Case Study in Punjab", *Economic and Political Weekly* 52(30): A5367-5375. December 30.

Little, P.D. and M.J. Watts (eds.) (1994). *Living Under Contract: Contract Farming and Agrarian Transformation in Sub-Saharan Africa*, Madison. University of Wisconsin Press. pp.216-47.

Mayers, J. and S. Vermeulen (2002). *Company—Community Forestry Partnerships: From Raw Deals to Mutual Gains?*. London: International Institute for Environment and Development (IIED).

MFPI (1998). *Food Processing Industries: Outline of Assistance under Schemes of the Ninth Plan 1997/2002*. Government of India, November.

MoAC (Ministry of Agriculture and Cooperatives) (2002). "Linking Production and Marketing under Contracts", Approach Paper for the Workshop (Contract Farming) held on February 8 (translated from Thai).

Page, S. and R. Slater (2003). "Small Producer Participation in Global Food Systems: Policy Opportunities and Constraints", *Development Policy Review* 21(5-6): 641-654.

Pingali, P. and Y. Khwaja (2004). "Globalisation of Indian Diets and the Transformation of Food Supply Systems", *Indian Journal of Agricultural Marketing* 18(1): 26-49.

Reardon, T. and J.A. Berdegue (2002). "The Rapid Rise of Supermarkets in Latin America: Challenges and Opportunities for Development", *Development Policy Review* 20(4): 371-388.

Sartorius, K. and J. Kirsten (2005). "The Boundaries of the Firm: Why do Sugar Producers Outsource Sugarcane Production?", *Management Accounting Research* 16(1): 81-99.

Schwentesium, R. and M.A. Gomez (2002). "Supermarkets in Mexico: Impacts on Horticulture Systems", *Development Policy Review* 20 (4): 487-502.

Scott, C.D. (1984). "Transnational Corporations and Asymmetries in the Latin American Food System", *Bulletin of Latin American Research* 3(1): 63-80. January.

Shiva, V. and T. Crompton (1998). "Monopoly and Monoculture- Trends in Indian Seed Industry", *Economic and Political Weekly* 33(39): A137-A151.

Simmons, P., P. Winters and I. Patrick (2005). "An Analysis of Contract Farming in East Java, Bali, and Lombok, Indonesia", *Agricultural Economics* 33: 513-525.

Singh, G. (2005). "Contract Farming of Mint in Punjab", *Indian Journal of Agricultural Marketing* 19(2): 121-129.

Singh, G. and S.R. Asokan (2005). *Contract Farming in India- Text and Cases*. New Delhi: Oxford & IBH.

Singh, Roopam (2008). "Effectiveness of Contract Farming Practices for Agricultural Development and Equity: A Case of Hoshiarpur District in Punjab". New Delhi: Oxfam International. (Unpublished Report).

Singh, S. (2000). "Contract Farming for Agricultural Diversification in the Indian Punjab: A Study of Performance and Problems", *Indian Journal of Agricultural Economics* 56(3): 283-294.

———. (2002). "Contracting Out Solutions: Political Economy of Contract Farming in the Indian Punjab", *World Development* 30(9): 1621-1638.

———. (2005a). *Political Economy of Contract Farming in India*. New Delhi: Allied.

———. (2005b). "Role of the State in Contract Farming in Thailand: Experience and Lessons", *ASEAN Economic Bulletin* 22(2): 217-228. August.

———. (2005c). "Contract Farming System in Thailand", *Economic and Political Weekly* 40(53): 5578-5586. December 31

———. (2008a). "Marketing Channels and their Implications for Smallholder Farmers in India", in E.B. McCullough, P.L. Pingali and K.G. Stamoulis (eds.), *The Transformation of Agri-food Systems: Globalisation, Supply Chains, and Smallholder Farmers*. London: Earthscan Press. Chapter 14, pp.279-310.

———. (2008b). "Contract Farming for Agricultural Development: Review of Theory and Practice with Special Reference to India", in S.K. Bhaumik (ed.), *Reforming Indian Agriculture: Towards Employment Generation and Poverty Reduction*. New Delhi: Sage (in honour of Prof. G.K. Chadha). Chapter 8, pp.165-195.

———. (2008c). "Producer Companies as New Generation Co-operatives", *Economic and Political Weekly* 43(20): 22-24, May 17.

———. (2009). *Organic Produce Supply Chains in India-Organisation and Governance*. New Delhi: Allied. (Forthcoming).

Warning, M., N. Key and W.S. Hoo (2003). "Small Farmer Participation in Contract Farming", A Draft Paper.

Weatherspoon, D., J. Cacho and R. Christy (2001)." Linking Globalization, Economic Growth and Poverty: Impacts of Agribusiness Strategies on Sub-Saharan Africa", *American Journal of Agricultural Economics* 83(3): 722-29.

Welsh, R. (1997). "Vertical Co-ordination, Producer Response, and the Locus of Control over Agricultural Production Decisions", *Rural Sociology* 62(4): 491-507, Winter.

White, B. (1997). "Agro-industry and Contract Farming in Upland Java", *The Journal of Peasant Studies* 24(3): 100-136.

Winson, A. (1990). "Capitalist Coordination of Agriculture: Food Processing Firms and Farming in Central Canada", *Rural Sociology* 55(3): 376-394, Fall.

Witsoe, J. (2006). "India's Second Green Revolution? The Sociological Implications of Corporate-led Agricultural Growth", in D. Kapur (ed.), *India in Transition: Economics and Politics of Change* Series. Philadelphia: Centre for the Advanced Study of India. Fall.

Wolf, S., B. Hueth and E. Ligon (2001). "Policing Mechanisms in Agricultural Contracts", *Rural Sociology* 66(3): 359-381.

7 Futures Trading and its Impact on Farmers

Select Case Studies

V. Shunmugam and
Danish A. Hashim[1]

With the role of governments in the markets shrinking and economic borders merging, farmers in remote rural areas with 'virtually no access to information' had very little say in the prices they receive to sustain themselves, in developing countries. Growing dependence of a large section of population on agriculture and increasing fragmentation of arable land, coupled with rising pressure on it, led to a spurt in incidence of poverty in these countries. And, this also added to the rising trend of social unrest.

With more than two-thirds of population dependent on agriculture and allied activities and the Indian government retracting from the administered market regime, policymakers felt a dire need for empowering farmers not merely to sustain themselves but to make them the stakeholders in the process of economic growth. Futures trading with a long history of 125 years certainly came in handy. It was realised that commodity derivative markets needed to be revived. Further, penetration of ICT which resulted in huge reduction of costs proved to the policymakers that national online commodity exchanges would fulfill the need for empowering the farmers both directly and indirectly to enable better planning of their farming activity and after having produced what is needed for the markets, to help them realise better prices by appropriately timing the marketing strategies. The Indian government announced the new agricultural policy during 2000 which became a watershed in the history of futures trading in the country wherein the intent to restart commodity

1. Views expressed are those of the authors and not of the exchange.

futures was reflected upon. Finally, this led to the launch of national online commodity exchanges during FY 2002-03.

Trading in national online multi-commodity exchanges grew organically since then, leading to the overall development of the commodities ecosystem as a result of their spill-over effects. As these beneficial changes went unnoticed, the inflation phenomenon came in with a heavy hand to crush commodity derivatives to make it go into oblivion, hiding the overall developmental benefits that futures trading brought in for the farming and farm produce market stakeholders. Hence, here is an attempt by us to study the select ecosystems of cardamom and mentha oil to find out what changes have really been reflected upon by the farmers and other stakeholders of these two crops grown in two distinct geographical parts of India.

The futures market plays a key role in the marketing of agricultural and non-agricultural commodities by providing industries and the farming community with a transparent price discovery platform that enables them also to hedge their price risk/volatility.

While forward contracts were the earliest form of commodity derivatives, futures contracts have also existed for centuries in one form or the other. The first record of organised futures trading dates back to 17th century Japan. The country's feudal landlords used to ship surplus rice to storage warehouses in the cities and then issue tickets promising future delivery of the rice. These rice tickets, which represented the right to take delivery of a certain quantity of rice at a future date at a specified price, were traded on the Dojima rice market near Osaka in 1730. In the US, the Chicago Board of Trade (CBOT) was the first organised futures market offering 'grain' contracts (beginning in 1848), with which futures trading became popular in the mid-19th century. Metals followed suit with contracts traded on the London Metal Exchange (LME) in 1878. Thereafter, a number of commodity exchanges facilitating futures trading in numerous agri- and non-agri commodities sprang up the world over.

In India, the earliest reference to 'futures' can be found in Kautilya's *Arthashastra*. Organised commodity derivatives trading began with the Cotton Trade Association's debut in futures in 1875. Cotton merchants of Bombay followed the US and the UK, and Cotton Contracts Committee was established by the government to regulate futures trading. But the Committee was soon (1919) replaced by Cotton Contract Board. Futures in oilseeds became organised with the inception of Gujarati Vyapari Mandali in 1900 in the then Bombay. And then the derivatives market in India developed in several other commodities over the years: raw jute and jute goods in Calcutta (1912), wheat in Hapur (1913) and then bullion in Bombay (1920).

However, widespread fears that derivatives trading was fuelling unnecessary speculation in essential commodities and, thus, were detrimental to healthy functioning of the markets for the underlying commodities and, therefore, also to farmers, drove the government to suspend the trade in several commodities one by one. First, the Government of Bombay barred options trading in cotton in 1939, followed, in 1943, by a ban on forward trading in oilseeds and also in commodities such as food grains, spices, vegetable oils, sugar and cloth.

Thereafter, amid the devastating effects of the Great Depression during 1939-1945, the British rulers imposed controls over the financial markets. Consequently, the Indian commodity futures market disintegrated into virtual extinction. In reality, even, in hibernation, it continued in the form of over-the-counter (OTC) contracts in negligible volumes. Almost a decade later, Parliament passed the Forward Contracts (Regulation) Act, 1952 (FCRA) to regulate commodity futures trading in the country. With reforms in its financial markets in the early 1990s amid rapid liberalisation and globalisation of the Indian economy, the Kabra committee, set up by the government in 1993 to examine the role of futures trading, made several recommendations including certain amendments to FCRA and strengthening of the Forward Markets Commission (FMC). Soon, the government allowed futures trading in all the commodities recommended, and the trade resurfaced after nearly four decades, as realisation that derivatives do perform a role in risk management dawned. FMC and the government, then on a fast-track mode, encouraged the idea of setting up national level electronic commodity exchanges with state-of-the-art infrastructure and global best practices, and thus the Multi Commodity Exchange of India Ltd. (MCX), the National Commodity and Derivatives Exchange Ltd. (NCDEX) and the National Multi-Commodity Exchange of India Ltd. (NMCE) came into being.

At present, 24 commodity futures exchanges, including 21 regional bourses and the three national bourses, are operational in India. The combined turnover of domestic commodities exchanges, what began with a notional value of Rs 1,29,364 crore in 2003-04 increased to Rs 36,77,226 crore in 2006-07. Following a ban on some commodities in January 2007 and then in May 2008, and imposition of higher margins and stringent norms for trading, the growth in trade volumes slowed down in 2007-08. But the commodity futures market staged a comeback in 2008-09 with a sharp increase in the turnover to Rs 52,48,956 crore. As the percentage of gross domestic product (GDP) at market prices, the total trade accounted for 97.3 per cent in 2006-07, which only marginally slipped to 94.1 per cent in 2007-08 but shot up to 106.4 per cent in 2008-09. In fact, the Indian commodity futures market has emerged as one of the fastest growing markets with a combined trade turnover of around Rs 52.48 trillion ($1.14 trillion), and the phenomenal growth (110 per cent compounded annual average growth since the market's resurrection in 2003). Given the growth in trading volumes and increasing integration of Indian economy with the rest of the world, the Indian commodity futures market has begun to be recognised among the top derivatives exchanges of the world.

Impact of Futures Trading on Farmers

Futures trading, theoretically speaking, enables farmers bargain for better prices from intermediaries and also helps them manage market risks by participating on the trading platform. Indirectly, it has also helped the farmer largely by improving his decision-making ability through an appropriate enabling environment. This happens indirectly through a number of ways starting from transparent futures prices, funding on warehouse receipts, better standardisation and quality testing infrastructure, increased quality awareness, etc.

Despite the direct and indirect benefits created by the commodity futures exchanges to empower the farming community, the debate on their failure to meet the objectives (for which they were instituted) has so far largely loomed around the inability of the farmers to participate in the exchange based trading and hence, the irrelevance of such a platform as an alternative way to receive assured prices. However, the indirect benefits that accrue to the farming community have not been much talked about. While a part of the benefits of the existence of exchange-traded derivatives can be proved by empirical analysis of the exchange-discovered prices and spot market prices such as by comparing and contrasting them with the help of statistical tools, these would remain largely unfelt by stakeholders and not be attributed by either the farmers or the traders in the ecosystem to the presence of futures in their respective commodities. Nor is there any statistical tool that can attribute the same to the existence of trading in commodity derivatives. Only a primary survey undertaken to study the impact of futures trading on the ecosystem and hence, the farmers can help attribute the benefits to the existence of derivatives trading. A survey was undertaken by the staff of our exchange's research department and the data and statements presented herein are based on the videographed interviews conducted during the study.

International Experiences

Before looking at the Indian markets that are barely five years old, it would be ideal to understand what various studies emphasise about the direct and indirect impact of futures on the farming community from across the world. A study of available literature in the area suggests that the existence of commodity exchanges impacts the livelihood of farmers growing crops traded on these platforms both directly and indirectly.

The following case studies explain the efficacy of exchange-traded hedging instruments in managing price risks. Futures contracts of agricultural products on various exchanges are being utilised by farming bodies and representatives to hedge their risks. In some cases, the coverage is more than 60 per cent of the production area. For instance, on Paris Bourse SA in France, contracts on rapeseed, introduced in 1994, have increasingly been used by cooperatives and processors and now cover 60 per cent of European production.

Guatemala: Managing Risks for Easy Access to Cost-Effective Finance

National Coffee Association (ANACAFE), a private non-profit organisation, introduced a coffee credit system during 1994 aiming to improve the access of coffee producers to organised lending by commercial banks. As part of the programme, ANACAFE made it mandatory for the coffee farmers to hedge their price risks and offered access to low-cost funds. As per ANACAFE's estimates, the farmers have saved over 10 per cent of the total loan amount due to lower interest rates offered by them (Christopher *et al.,* 1999). The role of ANACAFE here was limited to handholding the farmers to both risk management process and to access the commercial banks in terms of paper work, etc. The share of the coffee farmers who had hedged their produce increased from nil to about 20 per cent due to the handholding provided by ANACAFE and the attractiveness of the overall programme.

Canada: Failed Efforts to Make Cattle Farmers Cover their Risks

Farm Credit Corporation of Canada in collaboration with Cargill Investor Services Limited offered customised option contracts to cattle farmers (UNCTAD, 2002). The option pricing covered both the price risk and the forex risk that a cattle farmer may face in the markets. However, the product did not sell much, and the price risk management programme was discontinued due to lack of interest in taking it up among the cattle growers. This was despite Cargill's effort to customise the contract to a smaller lot size, taking into consideration the specific needs of the growers.

Mexico: Unsuccessful Farmers' Participation Programmes

Similarly, farmers' participation programmes implemented by the Mexican government and the United States Department of Agriculture (USDA) to help farmers manage their price risks met with a failure due to lack of understanding among the farmers about the need for hedging and the process of doing the same (Combe, 1997).

Unlike Guatemala, the other efforts highlighted above did not succeed due to lack of appropriate training to the implementing officials leading to failure of appropriate handholding of the farmers.

Producers can use the exchange-discovered prices as a guide in their decision on cropping pattern or the timing of his sales. This reduces the vulnerability of farmers to the so-called 'cobweb effect' of interseasonal price fluctuations. Futures prices also act as important signals in commercial decisions made by other stakeholders such as traders, importers and exporters. For example, transparency of grain futures prices, together with the acceptance of foreign grains for delivery through the Johannesburg Stock Exchange (JSE)/South African Futures Exchange (SAFEX), acted to cushion the supply shortages in South Africa which can significantly increase the prices of staple food crops.

It can be argued that participants in the commodity sector of the developing countries do not need to actively trade on an exchange to reap the benefits of its existence. The availability of published market information, improved storage and logistics infrastructure, transparent quality standards and cheaper and easily accessible sources of finance are positive externalities from exchange operations that can make a significant improvement to farmers' livelihoods, even if farmers remain largely oblivious to the exchange's trading functions.

Indirect Benefits of Commodity Futures: A Case Study of Mentha Oil

Mentha oil, used for its various pharmaceutical and aromatic properties, is derived from mint plant which is largely cultivated in the Southern Himalayan range in the states of Punjab, Himachal Pradesh, Uttarakhand, Uttar Pradesh, Haryana and Bihar. It is taken up as the third crop in many of the areas that follow the *rabi* crop. The leaves are harvested, dried and then steam-distilled to extract mentha oil. Then this is crystallised through the process of spray drying to obtain the final product—menthol crystals—meant for consumption. Typically, the mint growers steam-distilled the dried, harvested mint leaves on their own or contract-distilled it and sold the mentha oil largely to the village level intermediaries. The intermediaries bought it at a price determined by the quality assessed through the traditional method of judging the intensity of the smell emanating from the given liquid. They pooled small quantities procured from farmers and finally sold it to larger intermediaries at a better price arrived at based on reliable quality certification done through the modern gas liquid chromatography (GLC) method. This facility is available at major marketing centres that receive larger quantities of products. Finally, the traders sold off the mentha oil to the processing units for further crystallisation. There were effectively 2-4 levels of intermediation taking away about 7.5-15.0 per cent of the final value of the produce. While a large part of mentha crystals were used domestically, a majority of mentha trade depended on exports of mentha crystal/powder. US imports constitute a higher share of more than half of Indian exports of mentha oil. The oil also faced competition from the synthetic mentha oil industry both domestically and internationally. However, use of mentha oil has increased manifold in recent times due to its therapeutic and aromatic value in many of the fast moving consumer (FMC) goods and pharmaceutical preparations.

Table 7.1

Mentha Oil: Indirect Benefits of Futures Trading

Parameters	*Pre-Futures (2004-05)*	*Post-Futures (2007-08)*
Avg. export price	Rs 438.5 per kg	Rs 607 per kg
No. of terminals	Nil	300
Area	1,00,000 ha (2004)	3,00,000 ha
Production	12,000 mt (2004)	30000-32000 tonnes
Quality standard	5 GLC labs	40-45 GLC labs
Value chain margin	11-12%	7.5-10.5%

Source: Spices Board India, MCX Survey and Trade Sources.

The trade in the mentha markets largely remained restricted to one-on-one deals and the prices remained relevant to the respective deals which were largely opaque. Hence, mentha oil producers and consumers largely remained at the receiving end leading to low price realisation at the origin and higher prices for the end consumers. Hence, lack of efficient markets remained a long felt need to break the deadlock between the prices realised by the farmers and the prices paid by the end consumers. This would go a long way in improving the cost competitiveness of the mentha supply chain not only to increase its consumption but also to improve the mentha farming ecosystem by helping them receive better prices.

Launch of mentha futures during 2005 remained a historical landmark which brought in transparency into the ecosystem leading to empowerment of the farmers. With the slow penetration of the mentha futures prices and the exchange-polled prices into the ecosystem (currently there are about 300 terminals in the major mentha growing belt), the farmers claimed to have become wiser to receive prices which are benchmarked to exchange polled/traded prices. Some large farmers with just enough market knowledge have also started participating on the MCX platform thereby deriving the benefits of the exchange-traded prices, besides spreading a word about the exchange and its prices among the community. Indirectly, the traders who delivered on the platform also helped the farmers by buying their produce for delivery at prices nearer to the exchange-traded prices.

In addition, it had also been observed that awareness about and opportunities for quality testing increased over a period in time due to the standardised quality contracts of the exchange and increasing GLC facilities in the ecosystem. Increased awareness in turn improved the commercial viability of GLC labs leading to rising investment in GLC facilities and reduction in cost of getting their produce quality tested and certified. Selling quality tested and certified mentha oil also led to increased price realisation for the farmers.

The exchange's delivery warehouses located in the major growing areas had not only increased opportunities to scientifically store their produce but also to realise better value of it by storing it for future sales in the market and at the same time meet his immediate cash requirements. Thus, the exchange-traded futures led to an overall improvement in the physical markets of mentha farming ecosystem.

The Cardamom Case Study

Cardamom (*elettaria cardamomum*) is one of the most intriguing and distinguished spices, with its distinctive, floral aroma and unique, clean flavour. This 'queen of spices', a truly versatile one, is native to the Western Ghats in Southern India. Traditionally, cardamom planters would either take their harvest to auction centres or sell it through local traders largely as an ungraded lot. Cardamom is sold finally to respective consumer groups in three grades—7 mm and above (for exports), 6-7 mm (for domestic markets), and below 6 mm (for the flavour extraction industry)—with the price being directly proportional to the size. The planters mostly arrived at the auction centres with their ungraded cardamom lot, for which they received prices based on mutual agreements about its quality among the sellers and the buyers. In a typical supply chain for cardamom, there existed about 3-5 levels of intermediaries entailing a margin 15-20.5 per cent. The traders also earn the value differential between an ungraded lot and a graded one.

Table 7.2

Cardamom: Indirect Benefits of Futures Trading

Parameters	*Pre-Futures (2004-05)*	*Post-Futures (2007-08)*
Avg. export price	Rs 368 per kg	Rs 495 per kg
No. of terminals	Nil	60
Quality standard	Nil	4 Grading and standardising facilities has been setup
Value chain margin	15-20.5%	3-4% (for people trading on Commex)

Source: Spices Board India, MCX Survey and Trade Sources.

Let us see how a futures contract traded on MCX could bring about changes in the markets. Firstly, the delivery centre was shifted to a major marketing centre in the main producing area compared with earlier attempts to have delivery at Cochin. This has not only made the physical market participants understand the fundamentals of the market but also led to the modernisation of the dilapidated warehouse, which was taken on lease by National Bulk Holding Coroporation (NBHC)—a collateral management arm of the exchange—to modernise and to make it a foolproof scientific warehousing structure to store cardamom through the length of the derivatives suited to high moisture conditions of the growing area. This resulted in shifting other value addition activities back to the growing areas besides offering better price information to the farmers to empower them and protect themselves from exploitation by middlemen.

Price Awareness, Investment and Employment Generation

At the time of the launch of MCX cardamom contract, there were only two terminals in the region. At present, there are about 60 terminals not only disseminating price information but also providing employment opportunities for about 100 people to conduct business in these terminals or run them or make them serve the purpose for which they have been installed in the growing regions. In addition, looking at grading and standardisation as a less capital-intensive entrepreneurial activity, four new grading and standardising facilities have been set up by planters/ entrepreneurs in and around the exchange delivery centre. It had not only moved the urban based value addition activities back into the producing areas but had also offered the planters to reap the value out of the grading process. As a result, the unsorted cardamom sales at auction centres have gone down from about 95 per cent to the current level of about 60 per cent. As the premium paid on 7 mm cardamom traded on the exchange, our conversations with other planters revealed that it had motivated them to invest more in crop inputs and operations to increase the quality of their harvest.

Warehouse Receipts

As a scientific warehousing facility for enabling collateral management and deliveries on the exchange had been set up, it had also thrown open opportunities for funding against commodities to the planters by banks and non-banking financial companies (NBFCs) operating in the growing areas. More importantly, as the exchange warehouse has been located at the production centre, it enabled producers to store and sell their produce on the exchange platform. Prior to the launch of MCX cardamom futures, there were no scientific warehousing facilities at the growing areas and storage took place in far away urban trading centres mostly not accessible to small and medium planters. National Bulk Handling Corporation of India Ltd. (NBHC), which maintains the exchange delivery facility is also tied up with Federal Bank not only to fund warehouse receipts but also to fund positions taken by the planters/exporters/traders on the exchange platform to protect themselves against price fluctuations at a much cheaper rate and to relieve them off immediate cash requirements.

Benefits of Futures Trading to Exporters

The fact that MCX launched futures contracts in cardamom had the underlying close their business made the exporters benefit out of trading and taking delivery of cardamom cutting short the supply chain between the planters and exporters. During 2007-08, trading in about 70,493 tonnes of exportable-quality cardamom took place on the MCX platform compared to Indian exports of 500 tonnes during the same period. Deliveries on the exchange platform estimated about 140.5 tonnes constitute about 28.1 per cent of the total exports. In fact, the quality specifications of the underlying on the MCX cardamom futures became so popular to be called 'MCX quality'.

Market Inclusive Growth: Needs Research/Policy/Institutional Boost

The above case study of cardamom clearly shows that given the right research, policy and institutional environment, the growing futures markets and the new breed of electronic spot markets can help in inclusive development of Indian agricultural sector. While the national spot markets determine the prices for any given day, the futures markets determine the prices for the times to come to enable farmers to choose the crop and determine the time of sale as it is the case with their manufacturing counterparts. Both go hand in hand to strengthen any commodity ecosystem given the right policy reforms which would enable them to function efficiently. This would require effective penetration of ICT, market linkages with financial institutions, development of market infrastructure, etc. That could be accomplished through effective policy reforms and public-private participation. Further educational, functional and financial literacy at the grassroot level would need to be improved so that the producers and consumers shall be empowered to reap the direct/indirect benefits of market inclusive growth. Market inclusive growth would not only boost productivity and growth in the agricultural sector, which is the mainstay of the majority in Indian economy, it could also bring about 'growth with equity' in the long run—the major goal of our planning process.

Conclusion

Futures are often mistaken for the markets in which farmers can directly benefit from it. Very few would know it that not many farmers can participate in futures trading, and it's the institutional changes and improvement in the commodity ecosystem that would indirectly help the farmers of the respective commodities to benefit from the existence of futures trading. Firstly, it was clear with the transparency that the exchanges brought in respective commodities ecosystem, intermediaries who thrived on lack of information have started moving out of the value chain helping the growers realise a larger share of consumers' rupee. In addition, the improvement in the market infrastructure both in terms of scientific storage, quality awareness, quality testing and certification, grading and standardisation, etc., among the growers had indirectly enabled them to receive better prices. Scientifically managed warehouses, the warehouse receipts system along with the backend tie-ups with the commercial banks have enabled the farmers not only to see a better future price for their produce but also receive them by fulfilling their cash needs with cheaper source of funds. In a way, it empowered the farmers to reap higher prices for seasonal commodities that they produced. On the overall, commodity futures with appropriate specifications, better level of penetration into the ecosystem and efforts of the exchanges to reach out to the stakeholders have really helped in the process of taking the markets and the value addition activities that had earlier moved to the costlier urban areas, back into the rural areas. It would in a way bring the rural economies into the mainstream economic growth that the national economic policy planners have been longing for.

References

Christopher, Gilbert L., Panos Varangis and Jan ter Wengel (1999). *Industry Representational Organisations and Market Reforms in Selected Coffee Producing Countries*. Amsterdam. November. *http://www.feweb.vu.nl/ medewerkers/cgilbert/iro.pdf*

Combe, De M. Olivier (1997). *The Role of Farmers' Associations in Commodity Price Risk Management and Collateralised Commodity Finance*. UNCTAD. October 7.

Spices Board of India. *www.indianspicesboard*

UNCTAD (2002). *Farmers and Farmers' Associations in Developing Countries and their Use of Modern Financial Instruments*. UNCTAD/DITC/COM/35. UNCTAD Secretariat. January.

8 Price Discovery in Indian Commodity Futures Markets

An Empirical Analysis of Select Agricultural Commodities

C.S.C. Sekhar and Danish A. Hashim

One of the major arguments advanced against agricultural trade liberalisation is the higher volatility of agricultural prices in international markets and its subsequent transmission into the domestic markets (Nayyar and Sen, 1994; Chand and Jha, 2001; Sekhar, 2004). Futures trading is proposed in the literature as one of the efficient mechanisms of addressing this problem (World Bank-UNCTAD, 1999; Dasgupta, 2004). However, there are critiques of futures trading as an efficient price smoothing institution too. Futures trading has traditionally been criticised on the following grounds (Kabra, 2007).

i) causes excessive and unwarranted price fluctuations,

ii) benefits mainly the speculators who are capable of manipulating prices and

iii) works against both the producers as well as consumers and benefits only the traders.

The severe price rise witnessed in the Indian food markets in the recent past (2007-08) has sparked a renewed debate about the role of futures trading in this price rise. The present study attempts to address the issue in a limited way. The paper attempts to find the relationship, if any, between the movements in futures and spot prices of some select agricultural commodities. The price discovery mechanism in the futures market may help to unravel the causal mechanism in the commodity markets.

The main objectives of the study are the following:

a) to identify the long-run relationship between the spot and futures prices in the select commodity markets and

b) in the presence of such a long-run relationship, to identify the underlying causation.

The paper is organised as follows. After a brief introduction in Section I, Section II provides an overview of the evolution of futures trading in India and a brief review of the recent literature on futures trading. Section III provides the analytical framework of the study and data followed by a discussion of the empirical results of the study in Section IV. Section V provides the concluding remarks and policy implications.

Futures Trading in India and Some Recent Studies on Futures Markets

Futures contract is an agreement for the delivery of a commodity at a specified later date at a price decided at the time of entering into the contract. Therefore, a futures contract may also be regarded as a commodity bond, payable in terms of the commodity at the contract specified date. Futures contract is an improved version of the forward contract in two respects.

i) Futures contract allows options trading i.e., the right to sell the contract, to both the parties. The final transaction takes place between the ultimate buyer and the seller.

ii) The concept of a clearing house, by keeping a margin from both parties, insures against default by either party that may result in breach of contract.

Bombay Cotton Trading Association (BCTA) was probably the first organisation to start futures trading in India (Natu, 1962). BCTA was established by European traders in 1875 as a joint stock company. Thereafter, Bombay Cotton Exchange (BCE) was formed in 1890 as an independent institution followed by the establishment of Bombay Cotton Brokers Association in 1915. The Government of India set up a Cotton Contracts Committee in 1918 which was later replaced by Cotton Contract Board in 1919. Finally, in 1922 a Central Cotton Association was formed which subsequently became popular as East India Cotton Association Ltd.

Around the same time futures markets developed for various other commodities too. Oilseeds futures market was established in Bombay in 1900 followed by similar markets in Gujarat, Saurashtra and Punjab. The first futures market in wheat started functioning at Hapur in 1913 followed by markets in Punjab, Uttar Pradesh, Bombay and Calcutta. Futures trading in jute and hessian started in Calcutta in 1912.

Functioning of these markets faced several restraints in the 1960s due to the widespread suspicion of market manipulation and price distortion by few players (Pavaskar, 1970). Futures market operations in molasses were banned in 1963 followed by a ban in trading of oilseeds like groundnut, rapeseed & mustard in 1965. The price and distributional controls on cotton were lifted in 1967 but the futures market operations remained closed, mainly because of the opposition from cotton textile industry to futures trading in cotton.

Some Recent Literature on Futures Markets

Recent empirical literature on the US futures markets indicates that price is discovered in futures trading (Brenner and Kronner, 1995). Futures prices are shown to lead spot (cash) prices for a number of agricultural commodities using cointegration and error correction models. It is shown that the nearby futures price has a much larger influence on the spot price although, at any particular point in time, futures contracts exist for a variety of delivery dates stretching a year or more into the future. Greater the time lag between contract and delivery, greater is the degree of independence between spot and futures prices (Leuthold, 1979; Naik and Leuthold, 1988; Bessler and Covey, 1991). Kawai (1983) analysed the effects of futures trading on price formation in the spot market using a stochastic rational expectations model. Lapan *et al.* (1991) analysed the production and hedging decisions of a competitive firm operating in forward and options markets. Other important studies relating to futures trading in agricultural commodities are Fortenbery and Zapata (1993, 1997), Covey and Bessler (1995) and Yang *et al.* (2001).

In Indian context, there is very little literature (Dasgupta, 2004; Kabra, 2007) on the role of futures markets subsequent to their beginning of operations in 2003. The present study is an attempt to fill this gap by particularly focussing on the commodities that are important in food consumption basket in the country.

Analytical Framework and Data

The efficient market hypothesis asserts that the forward (or futures) price of an asset/commodity should equal the expected value of the spot price in future. In other words, the one-period forward price is equal to the expectation of the spot price in the next period.

Let us denote the one-period forward price of the commodity, for delivery at time t, by f_{t-1} and spot price in time t by s_t.

$$E_{t-1}s_t = f_{t-1} \quad \text{...(1)}$$

If this relationship fails, speculators can expect to make pure profits from their trade in the concerned commodity market. If the participating agent's expectations are rational, then the forecast error for the spot price in t should have a conditional mean equal to zero. In other words,

$$s_t - E_{t-1}s_t = \varepsilon_t \text{ where } E(\varepsilon_t) = 0 \quad \text{...(2)}$$

Substituting (1) in (2) we get

$$s_t = f_{t-1} + \varepsilon_t \quad \text{...(3)}$$

Since s_t and f_{t-1} are usually non-stationary I(1) variables, this requires that, for ε_t to be stationary, the linear combination given by (3) must be stationary. Therefore, the unbiased forward market hypothesis states that there is a linear combination of futures and spot prices that is stationary. Such a stationary linear combination is called the cointegrating relationship.

We have used the cointegration and error correction models (ECM) to explore the relationship among spot, nearby and distant futures prices. It is reasonable to expect that the prices of spot, nearby futures and distant futures prices are not completely independent of each other since we are studying the behaviour of actual or expected prices of the same commodity at different points in time. Similar economic forces are expected to influence each of these prices although the individual movements may not be exactly identical. Therefore, we hypothesise that there exists a long-run relationship among these variables. More formally, we hypothesise that the three prices, namely, spot, nearby and distant futures prices are cointegrated. This hypothesis is then tested using formal econometric techniques.

Cointegration implies the following:

i) Although the individual variables are all non-stationary, there exists a linear combination of the variables (long-run equilibrium relation) which is stationary.

ii) Long-run equilibrium relation implies that the stochastic trends of the variables are linked and the variables cannot wander off completely independent of each other.

iii) Dynamic paths, or in other words, the current movements in each of the variables are linked to the deviation from the long-run equilibrium path (ECM).

Intuitively error corection mechanism may be explained as follows. If two non-stationary I(1) variables x_t, y_t are related such that the difference between them is a constant say 'd', then the cointegrating relation is denoted as given below.

$x_t - y_t = d \ (d>0)$

If in a particular period p, the distance between x_t, y_t increases to d^* ($d^*>d$), then in the subsequent period p+1, the following adjustment mechanisms are possible.

a) x_{p+1} may decrease and y_{p+1} may increase making $d^* = d$.

b) If x_p and y_p were both increasing, then increase in x_{p+1} may be much slower than that in y_{p+1} making $d^* = d$.

c) If x_p and y_p were both decreasing, then decrease in x_{p+1} may be much faster than that in y_{p+1} making $d^* = d$.

It is also possible that the entire adjustment is made by only one variable, that is:

d) No change in x_{p+1} but y_{p+1} may increase by (d^*-d) re-establishing the equilibrium.

e) No change y_{p+1} but x_{p+1} decreases by (d^*-d) re-establishing the equilibrium.

Which one of these adjustments is actually realised and to what degree is identified by the ECM.

We used the following ECM in the present study,

$$\Delta X_t = \pi x_{t-p} + \sum_{i=1}^{p-1} \pi_i \Delta X_{t-i} + \varepsilon_t$$

where Δ is the difference operator; $\Delta X_t = X_t - X_{t-1}$

X_t is a 3×1 vector given by

$$\begin{bmatrix} SP \\ NF \\ DF \end{bmatrix}$$

SP=spot price, *NF*=nearby futures price and *DF*=distant futures Price; π is a $3{\times}3$ matrix of coefficients relating lagged levels (not changes) in prices (x_{t-p}) to current changes in prices (ΔX_t); $\pi_i 3 \times 3$ matrix of coefficients relating lagged changes in prices (ΔX_{t-i}) to current changes in prices $(\Delta X_t), \varepsilon_t \sim \text{i.i.d}(0, \Sigma)$.

If there are r cointegrating vectors, we can factorise π into α and β such that:

$$\pi = \alpha_{n \times r} \beta_{r \times n}$$

$\alpha_{n \times r}$ is a $r \times n$ matrix of speed of adjustment parameters denoting the coefficients with which each of the r cointegrating vectors enter each of the n equations of the ECM, and $\beta_{r \times n}$ is a $r \times n$ matrix of cointegrating parameters.

The long-run structure is understood by testing the hypotheses on the parameters of the cointegrating vector β. The short-run dynamics are discerned by testing hypotheses about α and π_i (Johansen and Juselius, 1994). The λ_{trace} test is used to test for the number of eigenvalues of π that are statistically significant from zero. This number gives the rank r of π. The rank of π is equal to the number of cointegrating vectors.

Suppose there is a single cointegraing vector $\beta = (1, \beta_2, \beta_3)$ normalised w.r.t. SP_t. Let the cointegrating equation be $SP_t = \beta_1 NF_t + \beta_2 DF_t$. Our ECM with two lags in levels may then be written as:

$$\Delta SP_t = \alpha_1 (SP_t - \beta_1 NF_t + \beta_2 DF_t) + \pi_{11} \Delta SP_{t-1} + \pi_{12} \Delta NF_{t-1} + \pi_{13} \Delta DF_{t-1} + \varepsilon_{1t}$$

$$\Delta NF_t = \alpha_2 (SP_t - \beta_1 NF_t + \beta_2 DF_t) + \pi_{21} \Delta SP_{t-1} + \pi_{22} \Delta NF_{t-1} + \pi_{23} \Delta DF_{t-1} + \varepsilon_{2t}$$

$$\Delta DF_t = \alpha_3 (SP_t - \beta_1 NF_t + \beta_2 DF_t) + \pi_{31} \Delta SP_{t-1} + \pi_{32} \Delta NF_{t-1} + \pi_{33} \Delta DF_{t-1} + \varepsilon_{3t}$$

The hypothesis that the near futures price 'causes' spot price may be tested by using Granger Causality test in the following way. In the spot price equation of the ECM, if the coefficients of α_1 and π_{12} are significantly different from zero and in the near futures price equation the coefficients of α_2, π_{21} and π_{23} and are not significantly different from zero, then it may be concluded that the near futures price Granger causes spot prices and by implication, leads in the price formation mechanism in the spot market. In other words, causation needs to be evident only in a single direction i.e., from NF_t to SP_t.

Data

The time series data on spot prices, nearby futures prices and the distant futures prices are used in our analysis. The nearby futures price series (NF) is created by taking the nearest futures month to the delivery date and the distant futures price series (DF) is created by taking the month before NF. The data are drawn from the database of the National Commodity and Derivatives Exchange Ltd. (NCDEX). Daily closing price data spanning over a period from November 2005 to June 2007 has been used. The time period has not been extended beyond 2007 as the futures market operations of some of the commodities turned controversial due to their suspected role in the then price rise and were subsequently banned. The analysis is carried out for three commodities, namely, wheat, *chana* and soya.

Empirical Results

The detailed results for the three commodities are presented below. All the price series in each of the three commodity markets are found to be I(1) upon testing for non-stationarity using the ADF and PP tests.

Wheat

The λ_{trace} statistic indicates that there is one cointegrating vector in the space spanned by the vector of three prices—spot (SP), nearby futures (NF) and the distant futures (DF) in this market (Table 8.1). However, the t-statistics for the coefficients[1] of the cointegrating vector (normalised with respect to SP) are not significant.

Therefore, we tested for cointegration between SP, NF and SP,DF separately. The λ_{trace} results indicate that there is a single cointegrating vector in both the cases. Normalising with respect to SP, we find that in the first case, the t-value of NF is significant indicating presence of long-run equilibrium relationship between SP and NF. The estimation results of ECM for SP and NF, with equations ΔSP_t and ΔNF_t, show that the coefficients of VAR terms for ΔNF_{t-1} and ΔSP_{t-1} in the two equations respectively are not significantly different from zero (Table 8.2). However, the speed of adjustment coefficient is significant in both the equations showing that the variables are not exogeneous in the long run, although they appear to be exogeneous in the short run (insignificant VAR coefficients). Also, the results indicate that the null hypothesis of 'absence of Granger Causality' cannot be rejected for either variable. The ECM model for SP and DF also shows similar results indicating that the two variables are exogeneous in the short run but in the long run they are not exogeneous. Also, the null hypothesis cannot be rejected showing that neither variable is Granger causing the other variable.

It may, therefore, be concluded that the weak exogeneity of the two sets of variables (SP,NF in first cointegration and SP,DF in second cointegration) is rejected.

Chana

The λ_{trace} statistic indicates the presence of a single cointegrating vector among the variables SP, NF and DF (Table 8.1). Upon normalising with respect to SP, the coefficient of NF is of expected sign and significant while that of DF, although of expected sign, is insignificant. The estimation

1. The Johansen test is asymptotically equivalent to t-test in case of a single cointegrating vector. Since our sample size is sufficiently large (455), we can use this result.

Table 8.1

Johansen Test of Cointegration

S.No.	Variables	Test Assumption	λ_{trace}	No. of Cointegrating Equations (Normalised on SP)	NF_{t-1}	DF_{t-1}
Wheat						
1	SP,NF,DF	No trend in the data, Const. in CE	53.19**	1	-6.17	4.93
2	SP,NF	No trend in the data, Const. in CE	27.08**	1	-0.98**	-
3	SP,DF	No trend in the data, Const. in CE	25.78**	1	-	-0.94**
Chana						
1	SP,NF,DF	Trend in the data, Const. in CE	35.92**	1	-0.67**	-0.18
Soya						
1	SP,NF,DF	No trend in the data, Const. in CE	43.63**	1	4.66	-6.07
2	SP,NF	No trend in the data, Const. in CE	18.33*	1	-1.11**	-
3	SP,DF	No trend in the data, Const. in CE	19.50*	1	-	-1.15**

Note: * significant at 5 per cent level of significance.
** significant at 1 per cent level of significance.

of ECM shows interesting results (Table 8.2). In the ΔNF equation, the speed of adjustment as well as all the VAR coefficients are insignificant. Similar results are observed for ΔDF equation. However, in case of ΔSP equation, the speed of adjustment coefficient is significant but the VAR coefficients are all insignificant. This shows that SP is not exogeneous in the long run although there is exogeneity of SP in the short run. The same cannot be said about NF and DF. These variables appear to be exogeneous in the short run as well as long run, as judged from the significance of the parameter estimates. Therefore, there is some evidence for weak exogeneity of futures prices in the *chana* market and a consequent Granger Causation of the spot prices.

Table 8.2

Error Correction Models: Results

S.No.		*Variables*	α	ΔSP_{t-1}	ΔNF_{t-1}	ΔDF_{t-1}
Wheat						
1	SP,NF					
		ΔSP_t	-0.0488**	0.1386**	-0.0158	-
		ΔNF_t	0.039**	-0.0545	0.1007**	-
2	SP,DF					
		ΔSP_t	-0.0524**	0.1362**	-	-0.0069
		ΔDF_t	0.0358**	-0.0319	-	0.0710
Chana						
1	SP,NF,DF					
		ΔSP_t	0.0315**	0.0049	0.0255	-0.0513
		ΔNF_t	0.0314	-0.0438	0.0189	-0.0958
		ΔDF_t	0.0304	-0.0800	0.1126	-0.1225
Soya						
1	SP,NF					
		ΔSP_t	-0.0200	0.0374	-0.0068	-
		ΔNF_t	0.0473**	0.0089	-0.0235	-
2	SP,DF					
		ΔSP_t	-0.0428**	0.0413	-	-0.0149
		ΔDF_t	0.0279**	-0.0066	-	0.0138

Note: * denotes significance at 5% level

** denotes significance at 1% level.

Soyabeans

The λ_{trace} statistic indicates the presence of one cointegrating vector among the variables SP, NF and DF (Table 8.1). However, the parameter estimates of the cointegrating vector are not statistically significant. Therefore, we tried to explore the long-run relationship between SP, NF and SP, DF separately. The results in the first case show that there exists a long-run equilibrium relationship between SP and NF. The parameter estimate of NF in the normalised cointegrating vector is significant (at 1 per cent level). The ECM estimation results show that the speed of adjustment coefficient in ΔNF equation is significant while in the case of ΔSP equation it is insignificant (Table 8.2). The VAR coefficients in either equation are insignificant. This implies that SP is exogeneous in the long run as well as short run while NF is exogeneous only in the short run but not in the long run. The λ_{trace} statistic indicates the presence of a cointegrating vector between SP and DF also. The parameter estimate of DF in the normalised cointegrating vector is significant (the ECM results in this case are slightly different). The speed of adjustment coefficient is statistically significant in both ΔSP and ΔDF equations. However, none of the VAR coefficients is significant in either equation. Therefore, it may be seen from this set of results that neither variable is exogeneous in the long-run although both appear to be exogeneous in the short run.

Overall, it may be said that in soya market, there is short-run exogeneity among the variables but there is little evidence of exogeneity in the long run.

Conclusions and Policy Implications

Our analysis shows no evidence for weak exogeneity of futures prices in the wheat and soya markets. There is some evidence of exogeneity of futures prices in the *chana* market. Our results do not lend support to the hypothesis that futures prices are leading to a rise in spot price, at least in the wheat and soya markets although some effect of futures price on spot market cannot be ruled out in the *chana* market. Further research is needed, particularly on the linkages between producers'/traders' inventory decisions and futures prices, and the consequent effect on spot prices.

References

Bessler, D.A. and T. Covey (1991). "Cointegration: Some Results on Cattle Prices", *The Journal of Futures Markets* 11(4): 461-474.

Brenner, R.J. and K.F. Kroner (1995). "Arbitrage, Cointegration and Testing the Unbiasedness Hypothesis in Financial Markets", *Journal of Financial and Quantitative Analysis* 30: 23-42.

Chand, Ramesh, Dayanatha Jha (2001). *Trade Liberalisation: Agricultural Prices and Net Social Welfare' in Indian Agricultural Policy at the Cross-Roads*, S.S. Acharya and D.P. Chaudhari (eds.). New Delhi: Rawat Publishers.

Covey, T. and D.A. Bessler (1995). "Asset Storability and the Informational Content in Inter-Temporal Prices", *Journal of Empirical Finance* 2: 103-116.

Dasgupta, Basab (2004). "Role of Commodity Futures Markets in Spot Price Stabilisation, Production and Inventory Decisions with Reference to India", *Indian Economic Review* XXXIX(2): 315-325.

Fortenbery, T. and H. Zapata (1993). "An Examination of Cointegration Relations Between Futures and Local Grain Markets", *Journal of Futures Markets* 13: 921-932.

———. (1997). "An Evaluation of the Price Linkages Between Futures and Cash Markets for Cheddar Cheese", *Journal of Futures Markets* 17: 279-301.

Johansen, S. and K. Juselius (1994). "Identification of the Long-Run and Short Run Structure: An Application to the ISLM Model", *Journal of Econometrics* 63: 7-36.

Kabra, Kamal Nayan (2007). "Commodity Futures in India", *Economic and Political Weekly*: 1163-1170, March 31.

Kawai, Masahiro (1983). "Spot and Futures Prices of Non-Storable Commodities under Rational Expectations", *Quarterly Journal of Economics* 98(2), May.

Lapan, Harvey, Giancarlo Moschini and Steven D. Hanson (1991). "Production, Hedging and Speculative Decisions with Options and Futures Markets", *American Journal of Agricultural Economics*, February.

Leuthold, R.M. (1979). "An Analysis of Futures-Cash Price Basis for Live Beef Cattle", *North Central Journal of Agricultural Economics* 1: 47-52

Naik, G. and R.M. Leuthold (1988). "Cash and Futures Price Relationships for non-Storable Commodities: An Empirical Analysis Using a General Theory", *Western Journal of Agricultural Economics* 13: 327-338.

Natu, W.R. (1962). *Regulation of Forward Markets*. Bombay: Asia Publishing House.

Nayyar, Deepak and Abhijit Sen (1994). "International Trade and Agricultural Sector in India", in G.S. Bhalla (ed.), *Economic Liberalisation and Indian Agriculture.*

Pavaskar, M.G. (1970). "Futures Trading and Price Variations", *Economic and Political Weekly*, February.

Sekhar, C.S.C. (2004). "Agricultural Price Volatility in International and Indian Markets", *Economic and Political Weekly* XXXIX(43): 4729-4736. October 23-39.

World Bank-UNCTAD (1999). *Managing Price Risks in India's Liberalised Agriculture: Can Futures Markets Help?*. New Delhi: Allied Publishers.

Yang, Jian, D.A. Bessler and D. Leatham (2001). "Asset Storability and Price Discovery in Commodity Futures Markets", *Journal of Futures Markets* 21: 279-300.

9 Current State of WTO Negotiations on Agreement on Agriculture

Implications for India

Ramesh Chand

Background

The international community is engaged in discussing and framing rules for international trade and related aspects in agriculture, non-agriculture and services sectors since 2001 under Doha round of World Trade Organization (WTO) negotiations. During this seven years period the negotiations have passed through several phases but could not reach final agreement. The modalities on the Agreement on Agriculture (AoA) covered three elements *viz.*, domestic support, market access and export subsidies/ competition. Special treatment for developing countries was treated as an integral part of all the three pillars of AoA and non-trade concerns were also included. Despite several meetings and sessions held to come out with a consensus draft, positions taken by different members remain wide apart. The differences prevail over almost all areas like various boxes of domestic support, reduction in domestic support, tariff, tariff quotas, export subsidies and credits.

In the meantime several realignments took place among members and various positions are by and large represented by following: (i) G-20 group of developing countries that include Brazil, India, South Africa and Pakistan, (ii) G-33 group led by Indonesia including Pakistan and Sri Lanka and pressing for special and differential treatment for developing countries, (iii) G-10 group mainly developed countries including Switzerland, Norway, Japan, (iv) G-90 group comprising small, poor countries, least developed countries (LDC), the African, Carribean and Pacific group of states (ACP) and African groups (v) the USA and (vi) the EU. There have also been interactions and discussions within countries, between countries and between capitals and the Geneva process. While serious difference persisted among members, WTO headquarters has continued its efforts to strike a common ground.

The reasons for the delay in concluding the new round are the sharp differences among members on various aspects of AoA. Implementation of Uruguay Round (UR) commitments has been a tough task for several member countries and it has exposed the vulnerability of various segments of agriculture to global market forces. In most of the cases, expectations placed in UR AoA or promises related to this did not materialise. There is a widespread view that the UR was a disappointment (Chand, 2002; Grimwade, 2004). The promise was that trade liberalisation and implementation of AoA would bring large benefits to developing countries through improved access to developed countries markets, increased trade and better pricing environment for sharp tropical and other products of interest to developing countries. However, there was a distinction between reality and the promise. The biggest challenge to developing countries agriculture in the post-WTO period was posed by unprecedented and unforeseen decline in international agricultural prices. Because of this, developing countries' exports were badly hit and several countries were taken aback by import influx of those commodities in which they thought they had a strong competitive edge. This caused adverse impact on farmers' income and employment. Developed countries could safeguard their agriculture against low global prices by providing huge support to their farmers but developing countries neither had the mechanism nor resources to protect their agriculture and farmers against such adverse trading and pricing environment. The entire blame for this outcome is put on UR AoA. There is a feeling that developing countries did not bargain properly at the Uruguay Round and developed countries secured the balance of AoA in their favour. Besides being discriminatory, the agreement is said to be ably manipulated by developed countries to benefit their agriculture at the cost of developing countries. Such threats and fears are further reinforced by the World Bank study (2002) which predicts that the greatest beneficiaries of Doha Round of WTO negotiations would be the rich economies of Western

Europe. This time the member countries have naturally turned highly conscious and are very careful about the minute details of various provisions of any future AoA and they are trying hard to protect their interest adequately (Finger, 2001; Finger and Nogues, 2002). This has led to hardening of positions particularly relating to high level of domestic support and export subsidies in OECD countries, access to developed countries' market and special and differential treatment.

A recent development in taking Doha round further is that that WTO had called a mini ministerial meeting attended by 30 countries in the third week of July to finalise modalities for concluding the round by end 2008. The discussion continued for about eight days but the consensus could not be reached. Finally, on July 29, 2008 WTO declared that the negotiations have collapsed. The present paper discusses the recent (current) proposals, referred to as 'July 2008 Package' on various aspects of AoA on which WTO sought consensus of the member. This is followed by an assessment of proposals on the negotiating table from a developing country perspective. The lasts section explores possibilities of regional cooperation in agriculture particularly in relation to success/failure of negotiations on agriculture.

The July Package

The latest revised draft modalities prepared by Ambassador Crawford Falconer, chairperson of the agriculture negotiations, have been circulated as a sort of blueprint for the final deal. The draft negotiating text contains formulas for cutting tariffs and trade-distorting subsidies and related provisions. The draft has been discussed for eight days in a mini ministerial meeting from July 21 to 28; and WTO would prepare a revised draft taking into account the convergence and difference into account for further negotiations. Important features of the draft text are discussed below. The draft provides for special and differential treatment for developing countries under each element of AoA and it also provides for separate overall special and differential (S&D) treatment for agriculture in developing countries.

Domestic Support

Following the Hong Kong Declaration 2005, the proposal is to express all kind of trade distorting support as 'Base-level Overall Trade Distorting Support' which is taken as sum of:

(a) Final Bound Total Aggregate Measure of Support (AMS),

(b) De Minimis support (10 per cent of value of agricultural output during 1995-2000 for developed countries and 20 per cent for developing countries) and

(c) Higher of Blue Box payment as notified to the Committee on Agriculture or 5 per cent of value of agricultural production during 1995-2000.

A tiered formula is proposed to reduce Overall Trade Distorting Support (OTDS). For this, OTDS for developed countries is divided in three bands consisting of threshold support in the range of: (a) US $0-10 billion, (b) $10-60 billion and (c) more than $60 billion. Respective cuts in these bands are proposed to be between 50-60 per cent for (a), 66-73 per cent for (b) and 75-85 per cent for (c). According to these bands, EU comes in highest band (c), USA and Japan comes in middle band (b) and all other developed countries in band (a).

Table 9.1

Matrix Showing Bands and Cuts in OTDS and AMS

Particular	*Tier/Band*	*Threshold US$*	*Cuts*
a. OTDS	1	0-10	50-70%
	2	10-60	66-73%
	3	> 60	75-80%
b. AMS	1	0-15	45%
	2	15-40	60%
	3	> 40	70%

Source: World Trade Organization (2008).

As it is obvious, the level of threshold range is not related to the size of agriculture in a member country. Thus, there is possibility of a smaller size country even with higher support to fall in lower tier as compared to a country with larger size of agriculture and relatively less support. In such cases the package seeks complementing tier related reduction with additional criteria.

Developed country members with high relative levels of base OTDS in the second tier (i.e., at least 40 per cent of the average total value of agricultural production in the 1995-2000 period) shall undertake an additional effort. The additional reduction to be undertaken shall be equal to one-half of the difference between the reduction rates specified in tier 1 and 2 above. Similarly, developed country member with high relative level of final bound total AMS shall undertake additional cut.

Members in the first two tiers shall reduce base OTDS by one-third on first day of implementation and the remaining reduction shall be done in five equal steps. For members in the third tier, these rates are 25 per cent followed by five equal steps.

There is S&D treatment for developing countries. These countries with no Final Bound Total AMS (FBT AMS) do not have to reduce base OTDS. Other developing countries would reduce base OTDS by two-third of relevant rate for lowest tier for developed countries. Further, developing country will reduce OTDS in eight equal steps in eight years with 20 per cent reduction on the first day.

Similarly, developed country members shall implement reduction in AMS in six steps over five years while developing country would reduce in eight years. Developing countrics with AMS below under million would not be required to take any reduction.

De Minimis

De minimus level of Uruguay Round is to be reduced by no less than 50 per cent on the first day of implementation period. Developing country shall reduce De Minimis support by two-third of the reduction rate applicable to developed country and the time period would be three years.

Blue Box

There is maximum limit on Blue Box and it should not exceed 2.5 per cent of total value of agricultural production in the base period. There are also product-specific limits.

Green Box

July package describes Green Box in detail. This includes items like government services, public stockholding for food security, disaster related payments, insurance payments etc. The document seems to be putting lot of restrictions on each country for making payment for damage, loss to crops and animals due to natural calamities, pests and diseases.

Case of Cotton

In order to address the plight of cotton farmers in developing countries, cotton is taken up separately in the AoA. The proposed package calls for elimination of all forms of export subsidies on cotton by developed countries. AMS for cotton is proposed to be reduced according to following formula

$$RC = [RG + (100\text{-}Rg)^{*}\ 100] / 3^{*}\ Rg$$

Where:

Rc = Specific reduction applicable to cotton as a percentage.

Rg = General reduction in AMS as a percentage.

There is S&D treatment for developing country in the case of cotton also.

Market Access

There are four aspects of market access i.e., tariff cut, sensitive product, special product and special safeguard mechanism. At present tariff on some of the lines is imposed on per unit basis rather than on value of product. This poses serious difficulty in putting tariff lines in tariff bands. The proposal establishes method for calculating ad-valorem equivalent (AVE) where tariffs are not computed on value, except in the case of sugar. This creates a basis for putting tariff lines into bands for tariff cuts. There is considerable convergence on adopting linear cuts within the band and disagreement on level of cut is reduced considerably as can be seen from the Table 9.2 below.

Table 9.2

Tariff Lines Proposed

Band	*Developed Countries*		*Developing Countries*	
	Threshold (%)	*Range of Cut (%)*	*Threshold (%)*	*Range of Cut (%)*
1	0-20	50	0-30	
2	20-50	57	30-80	2/3rd of developed country in the corresponding range
3	>50	64	>80	
4	> 75	66-73	> 130	

Source: World Trade Organization (2008).

The bound tariffs are required to be reduced in five years by developed countries and in eight years by developing countries.

The July package provides for a sort of special and differential treatment for developing country members in two ways. One, developing country members have the flexibility to designate an appropriate number of lines as special products which are related to food security, livelihood security and rural development. Two, these countries will have the right to have recourse to a special safeguard mechanism based on quantity and price. This provision was earlier available only to 38 member countries.

Sensitive Products

Each developed country member will have the right to designate up to 4-6 per cent of tariff lines as 'sensitive products'. Where such members have more than 30 per cent of their tariff lines in the top band, they may increase the number of sensitive products by two per cent, subject also to the conditions outlined in paragraph 75 of WTO's report. Where application of this methodology would impose a disproportionate constraint in absolute number of tariff lines because tariff concessions are scheduled at the 6 digit level, the member concerned may increase its entitlement by that amount also (WTO, 2008).

Developing country member shall have the right to designate up to one-third more of tariff lines a 'sensitive products'.

Export Competition

Members are in agreement to ensure parallel elimination of all forms of export subsidies and discipline on all export measures having a sort of indirect affect like subsidies to be completed by the end of 2013 by developed countries and by 2016 by developing countries. Export subsidies are proposed to be reduced by 50 per cent by 2010.

While for developed countries progressive and parallel elimination of export subsidy is proposed, developing country members will continue to benefit from the provision of Article 9.4 of AoA for five years after end date for elimination of all forms of export subsidies.

Export subsidies on cotton are prohibited in the draft proposal.

Special and Differential (S&D) Treatment

Special Products

Developing country members shall be entitled to self-designate special products guided by indicators based on the criteria of food security, livelihood security and rural development. There shall be 10-18 per cent of tariff lines available for self-designation as special products. Up to 6 per cent of/no lines may have no cut. The overall average cut shall, in any case, be 10-14 per cent.

Special Safeguard Mechanism (SSM)

A price based and volume based SSM shall be available for all the tariff lines. Volume based SSM will consider import of preceding three years for volume trigger. Price based SSM will be applicable when CIF import price expressed in domestic currency falls below 85 per cent of average monthly MFN sourced price. The additional duty shall not exceed 85 per cent of the difference between the import price and the trigger price.

Fullest Liberalisation of Trade in Tropical and Diversification Products

The proposal gives a list of such products. It proposes reduction in tariff from scheduled tariff of 25 per cent to zero and 85 per cent reduction in higher tariff. There shall not be any sensitive product treatment for any of the product in the annexed list.

Assessment of July 2008 Package

Developing countries particularly G20 emerged as a strong group in articulating the agenda for WTO negotiations in Hong Kong round. Because of this, July Package of 2004 and Hong Kong Declaration (2006) got significantly changed and developing countries were able to push several of their proposals in the negotiations. It is now clear that because of very heavy domestic support in various forms and export subsidies in developed countries, agriculture in developing countries is not facing levelled field. Because of this, developed countries have to accommodate interest of developing countries and agree to different rules and obligations for this group. Despite full recognition of this aspect, consensus could not be reached on most of the issues. Member countries have to show lot of flexibility to reach final agreement. The flash point for collapse of mini-ministerial negotiations is reported to be disagreement on SSM and on reduction in support to cotton in USA.

The main achievement of negotiations of last eight years for developing countries is inclusion of S&D clause under each element of AoA and a separate S&D treatment clause for developing member countries of WTO.

There are some complications in respect of domestic support as the reduction is proposed based on absolute amount rather than considering domestic support relative to value of agriculture output. Similarly, separation of trade distorting domestic support and AMS for reduction commitment complicates the implementation. Further, when there is separate box for Blue Box support, the rationale of its inclusion in FBT AMS is not clear. The agreement institutionalise Blue Box to provide assistance upto 5 per cent of value of produce. The Green Box would also be there though the package seeks only non-trade distorting or minimal trade distorting measures in it. It needs to be noted that even in the UR agreement Green Box was defined only to include non or minimal trade distorting support but subsequently the box was used to shift Amber Box support to exempted Green Box. Therefore, Green Box measure must be defined very clearly to avoid any ambiguity. In the area of market access tariff reduction formula is yet to be decided. The framework permits developing countries to designate list of 'special products' which would not be subject to market access commitments. In the same breath, the package remains a sort of special safeguard mechanism for 'sensitive products' which would allow developed countries to deny market access in their countries to the products defined as sensitive.

Implications for India

Progress of WTO negotiations since Doha shows that most of the concerns raised by India have been addressed. There is S&D for developing countries in each element of proposal on AoA and there is additional S&D like special product and SSM for countries like India. July negotiations had moved very close to consensus among G-8 but small difference on trigger and subsidies in USA led to collapse of the talks. It seems there is enough flexibility in the proposed package to enable India to protect its agriculture against imports from developed countries. Moreover, the undercurrent supports that global prices of agricultural commodities would remain under pressure from demand side with little possibility of them going very low to require high protection. The delay in concluding the round is now going in favour of developed countries as they are getting more breathing time for reduction in subsidies and tariff. Indications are that the consensus emerging on AoA offers a good deal for India, and it is certainly better than the status quo.

References

Chand, Ramesh (2002). *WTO, Trade Liberalisation and Indian Agriculture*. New Delhi: Mittal Publications.

Finger, J. M. (2001). "Implementing the Uruguay Round Agreement: Problems for Developing Countries", *The World Economy* 24 (9): 1097-1108, September.

Finger, J.M. and J.J. Nogues (2002). "The Unbalanced Round Outcome: The New Areas in Future WTO Negotiations", *The World Economy* 25(3): 321-340, March.

Grimwade, Nigel (2004). "The GATT, the Doha Round and Developing Countries", in Homi Katrak and Roger Strange (ed.), *The WTO and Developing countries*. New York: Palgrave Macmillan.

World Bank (2002). *Global Economic Prospects and the Developing Countries 2002: Making Trade Work for the World's Poor*. Washington D.C.

World Trade Organization (2008). *Committee on Agriculture Special Session*. Report No. TN/AG/W/4/Rev.4. December 6.

10 Flood, Farmers and the Future

Nirmal Sengupta

As per the compilation found in an ADB study (Lin *et al.*, 2007) in the 100-year period ending 2004, a total of 9,632 natural disasters were reported from all over the world. Floods accounted for about 28 per cent of the total number and about 40 per cent of the economic damages due to these disasters. Asia was affected by floods more than any other continent. It experienced a total of 1,229 flood disasters, or 41 per cent of the world total during the 100 years. South Asia was the most affected region in Asia (and globally) with a reported incidence of 427 flood disasters. India was the most affected country in Asia. The Planning Commission (2002: 8.1.22) estimated that on an average 7.55 million hectare (mha) is affected annually by floods, of which 3.55 mha is cropped area. Average annual loss from floods includes 1,595 human lives, 94,772 heads of cattle and damage of 1.2 million houses.

How do we deal with this major source of disaster to agriculture needs to be discussed in a conference on the 'Future of Indian Agriculture'. How serious is the threat? Is it just one of the many natural calamities that Indian agriculture face? In case of another constant threat, the drought, there is no dispute that significant progress has been made in dealing with the problem. For flood threat, there is no clear answer. Do we have the technology to control flood, or institutions to deal with the threat? There are many confident claims, but no definite answer that survives scrutiny. This paper summarises our knowledge of India flood, its incidences, impact and policies, and thereby suggests future courses of action.

How Serious is the Threat ?

The Working Group on Water Resources of Eleventh Five Year Plan estimates (Planning Commission, 2007: 1.5) average annual damage at about Rs 1,805 crore. However, the damages reported for the recent years (*viz.*, Table 10.1) are much higher. This is expected. Damages depend on the levels of developments. But there is a far more important methodological question. The data series is in current prices. Most parts of the rapid increase over the years is not development but for price movement. On converting the damage statistics in constant prices the average annual damage is Rs 4,216 crore in 1999-2000 prices, i.e., about Rs 6,250 crore in 2007-08 prices. This compares well with the current damage data. The Eleventh Plan Working Group summarised the flood damages reported by the states and the relief fund released during the first three years of the Tenth Plan as:

Table 10.1

Flood Damage and Relief in Tenth Plan

(Rs crore)

Year	*Flood Damage Reported*	*Recommended Calamity Relief Fund*
2002-03	2575	1600
2003-04	4434	1587
2004-05	3337	1286

Source: Government of India, 2008.

How serious are the losses? The damage caused in India by the great 2004 tsunami was estimated at Rs 7,800 crore. The June 2008 flood in the upper Mississippi River Basin is one of the two worst floods in the US history. About 11 million people were affected, 20 people were killed and 5 million acres of agricultural land were inundated. The corresponding figures for average annual losses of Indian floods are uniformly higher: 33 million people affected, 1,589 people killed and 8.74 million acres of crop area damaged. Economic loss from the June 2008 flood of US is estimated at a whooping $10 billion, 10 times higher than the Indian estimates of average annual losses from floods of India. The gigantic difference is because US estimation methods use: (a) higher valuation of each component of damage and (b) includes damages to private property.

While releasing the flood data, the Central Water Commission (CWC) clearly indicates that the data of damages are to 'total crops, houses and public utilities'. The Commission does not equate it to total loss. But in course of processing it becomes so. The unaccounted part, due to the loss of private property in floods, may be very high. After the Mumbai flood in 2005, the General Insurance Corporation (GIC) alone had settled claims worth Rs 650 crore. After 2006 flood that hit several industrial belts like Vadodara, Hazira or Bhadrachalam, the Swiss Reinsurance estimated the total damage at Rs 15,000 crore or $ 3.39 billion (Swiss Re, 2007: 17). This is 30 times the estimate (Rs 507.09 crore) of the CWC.

Losses from other types of major catastrophes occur once in a while. But the recurrent floods of India are annual events. The following table shows the relative frequencies of major occurrences of different types of disasters in India over a 16-year period.

Table 10.2

Major Disasters in India 1990-2005

Type of Disaster	*No. of Disasters*
Cyclones (category IV-V)	4
Earthquakes (magnitude 6+)	5
Floods	Every year
Droughts	Every 2-3 years
Tsunami	1

Source: National Institute of Disaster Management.

It is evident from global data that both the frequencies and the magnitudes of damages from natural disasters like flood, hurricane etc., have increased several folds in later periods. The increased frequency might reflect the consequences of climate change. The rapidly increasing losses and the occurrences of several mega disasters in recent period call for another explanation. In general, damages depend on the levels of developments. But residents and agencies do not show much enthusiasm for investing in areas frequently visited by disasters. Thus, damages are much higher when a disaster strikes an area where it is less expected. After 2006 flood that badly hit Gujarat, the insurance claims mounted to Rs 1,800 crore (Swiss Re, 2007: 17).

It follows that one possible way of limiting damages is to prevent economic development of the area liable to disaster. As will be discussed later, this is a standard strategy used in US and elsewhere, particularly for flood damage mitigation. US regulations restrain developmental works in flood-prone areas. Similar provisions exist in other countries. In India, though there is a central guideline, very few states have taken up this step, known as 'flood zoning'. But even without the regulation there is little incentive to develop. On an average, an area of 7.55 mha is affected by floods every year. Total flood-prone area in the country, as per the *Report of Rashtriya Barh Aayog* (1980) was 40 mha. This has now been reassessed to 45.36 mha. It implies that on an average any particular part of the vast flood-prone area is likely to experience a flood in each six years. This is only an average. Some parts are visited by floods more frequently, some others are less. It is claimed that 18.22 mha of the flood-prone area has been reasonably protected by the end of Tenth Plan (Planning Commission, 2007: 5.1). If there is some truth in the claim then the frequency of floods in the unprotected areas are even more. In all likelihood, the residents and agencies may not have much incentive to invest in these areas which are likely to be ravaged by floods in quick succession. In other words, the social cost is far more than the actual damages. No other disaster visits with such frequencies. Therefore, heavier investments are made in hurricane or earthquake-prone areas. Then, when the disaster strikes once or twice in about a 100 years, damages are much higher. But that is rare and in small pockets. In contrast, over extensive flood-prone areas that experience repeated floods at short intervals, the development potential is sacrificed. Is their a future for these impoverished millions?

Approaches to Flood Management

Popular belief is that flood threat could be eliminated if the government was serious. This is only partly true. Following the spectacular advances with cement, mortar and concrete, in early 20th century it was hoped that floods would be relegated to history. The beginning of the Tennessee Valley Authority (TVA) was flaunted as ushering in of a new era of bygone flood menace. TVA inspired many similar projects, including the Damodar Valley Corporation (DVC) and the Hirakud dam in India. But soon the hope faded away. Despite having the TVA, flood continued to occur in Tennessee Valley. In February 1973, Chattanooga city was devastated by a 37 feet high flood. Parts of the TVA was designed primarily to protect this traditionally flood affected city. This was not the lone incidence of flood in post-project period. In just about 30 years, the TVA realised that its dams and reservoirs could only moderate, but would not prevent all flooding. Since then, TVA also tries to 'keep the people away from floods'.

In USA or in other countries, construction of dams primarily for flood control have ceased from being used. The primary policy towards flood management developed along a different course, after accepting the reality that floods are inevitable. Following a disaster the task that arises is of reconstruction. After some bad experiences, even the highly developed insurance sector of US refused to extend insurance for flood havoc, making reconstruction difficult. There was no other way out except the government stepping in, either by outright grants or by undertaking insurance services. US National Flood Insurance Program (NFIP) was created in 1968. Local communities may avail this insurance facility if they commit to abide by certain guidelines while investing in flood-prone areas. Essentially, these guidelines, enacted by the US Congress, are meant to limit development of flood-prone areas. In the UK, private insurance companies still provide flood insurance regardless of the risk level. French government has enacted a mandatory catastrophe insurance scheme. However, experiences show that this policy too has its limitations. In the developed world, insurance agencies have run into severe problems after the catastrophic disasters like Katrina hurricane. In 1984 China made an attempt to introduce insurance for a highly flood-prone areas, but did not succeed.

In conventional water resource management studies these different strategies of coping with flood are classified as structural and non-structural measures. The whole package includes many other measures, like early warning, or flood shelter. From the perspectives of economics and financial demands, however, control structures and insurances may be considered as the essential descriptions of structural and non-structural measures respectively. Recent development of a disaster management science provides a better framework for proper appreciation of flood policies. Although in everyday uses the two terms are used interchangeably, in the literature of disaster science a distinction is drawn between 'disaster' and 'hazard'. When a 'natural hazard' occurs in a vulnerable area it is a 'natural disaster'. One may describe:

Disaster = hazard × vulnerability

The distinction in terms is not just of theoretical interest. It results in sharp policy distinctions. While earthquakes and tsunamis are 'natural hazards' that cannot be eliminated, it is possible to reduce vulnerability by careful steps. Success of such policies may ultimately avert potential disasters, even though the hazardous event occurs. On the other hand a hazard like chemical pollution need to be managed not by reducing vulnerability of people likely to be affected, but by containing the production of the hazardous substance. Thus, appropriate policies differ from case to case. Epidemics for example, find immunisation as the best strategy in some cases, and eradication of the carrier in others.

Global experience of structural measures for controlling floods make it clear that like any other natural hazard, floods cannot be eliminated. Also, structures like large dams are discredited for various other reasons. As a result, flood management policies at present are primarily directed to reduce vulnerabilities, just as those other natural disasters. Unfortunately, an important distinction from other disasters tends to be forgotten. Unlike earthquakes and tsunamis, the intensities of flood hazards can be moderated by structural measures. In fact, in the case of Chattanooga flood mentioned earlier, dams and regulators of TVA succeeded in reducing the crest of flood water from 52.4 feet[1] to 37 feet, greatly reducing the devastation. Similar claims of moderation are made for other structures. If there is some truth then flood disaster management policies need to be directed towards reductions of both hazard and vulnerability.

Also, the right mix of the two types of strategies would differ from place to place. Although global flood policies have gradually moved from the primacy given to control structures in the 1930s to primacy of flood insurance measures, only the developed countries have attained some moderate success in flood insurance. In countries of South Asia or Eastern Asia even the possibility of extensive use of flood insurance is limited. The insurability of frequently flooded areas, abundantly found in the monsoon Asia, is doubtful. Also, viability of insurance schemes depend on making due arrangements for limiting development, hence potential loss. Asian countries cannot write off vast tracts from the map of development.

1. TVA website: *http://www.tva.gov/river/flood/index.htm*

Insurance programmes were relatively successful in the western countries where only small parts of the countries in question are flood-prone. But importing such policies here would not be effective, Asian countries need to develop their appropriate policies.

Evolution of Indian Flood Policy

As expected for the most flood-prone area in the world, flood control efforts in India have a very long history. Modern engineering interventions date back to 1850s, with the construction of the Damodar embankment. In just about 10 years it was clear that the structure did not make much difference. Since then the engineers in India were engaged in an intense and enduring debate on the appropriate technique of flood control. The colonial government was not prepared to meet the expenses of flood control. So the high quality technical debate remained merely a theoretical discussion, until the nationalist forces were about to share power. In 1938, the National Planning Committee under the chairmanship of Jawaharlal Nehru, began serious discussion about development of India. Flood control was included in its suggestions. Immediately after Independence, the constructions of DVC and Hirakud dams were undertaken initiating a massive structural effort for flood control. These were the flagship projects signalling the making of a new India.

In spite of such a start, flood control efforts through large dams soon came to an end. After the DVC dams, Hirakud on Mahanadi and Ukai on Tapi river, no other large dam was constructed primarily for flood control. Some other dams like Bhakra, Rengali or Gandhi Sagar were provided with flood cushioning and other moderation facilities. Flood control strategy has come to rely mostly on construction of embankments. The Eleventh Plan Working Group on Water Resources estimates (Planning Commission, 2007: 1.13.5) that a total area reasonably protected against flood by the end of the Tenth Plan is about 18.22 mha. That would be about 40 per cent of the estimated 45.36 mha total flood-prone area in the country. Along with dams and reservoirs this is suggestive of considerable progress in flood control works. Yet the data available for the period 1953-2005 does not confirm that there is any reduction in the incidence of flood (*viz.,* Figure 10.1).

Figure 10.1

India: Flood Affected Area

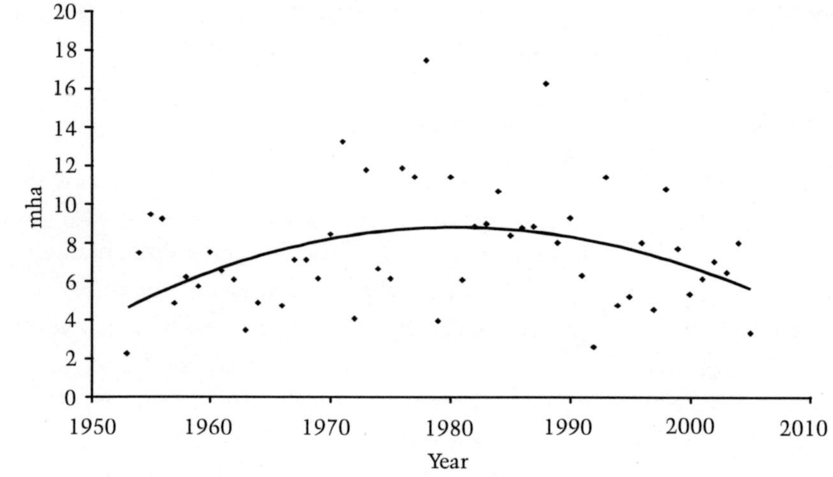

Source: Government of India (2006a).

The acreage affected by flood varies widely from year to year. A linear trend, however, shows that flood affected area is increasing throughout the Plan period at a rate of 20.8 thousand hectares (tha) per year. The data pertains to area affected and cannot be explained by the increase in property value over the years. Instead, it lends support to a claim that flood has been increasing over the years.[2] Figure 10.1 shows wide dispersion of yearly data. The non-linear fit may utmost support that the incidences were increasing in the earlier decades, which have been brought down in later years. By no means it lends support to success of flood control efforts in the plan period. However, the database may be considered insufficient for a definite conclusion about the increasing trend.

The Ninth Plan document (Part II, Table 4.2.11) includes a tell-tale table which shows that the expenditure for flood protection measures in the corresponding years was much less than the expenditure on relief measures (Table 10.3), which in turn, was much less than the estimated damages mentioned earlier.

2. This is also the global picture. Summarising the available evidence Lin *et al.* (2007) noted that 'countries that have constructed more dams do not necessarily have fewer flood disasters and economic losses.' In some cases data would rather support increase in flood after dam construction.

Table 10.3

Plan Expenditure and Flood Relief in Some Years

Year	*Plan Expenditure (Rs crore)*	*Relief Expenditure (Rs crore)*
1992-93	330.16	893.30
1993-94	364.19	564.00
1994-95	308.82	630.00
1995-96	348.67	729.40
1996-97	542.65	-

Source: Government of India (2006b).

Table 10.1 shows that this pattern has remained unchanged in more recent years. The demands, however, are much greater. After 2005 flood, the Chief Minister of Tamil Nadu sought Rs 1,742 crore from the National Calamity Contingency Fund (NCCF) for flood relief and rehabilitation measures.[3] Following the 2007 flood, the Chief Minister of Bihar filed an objection stating that the state was eligible to receive Rs 2,156 crore under Calamity Relief Fund (CRF), but the Centre assessed the requirement for relief at a meagre Rs 745.26 crore.[4]

Steps like flood plain zoning directed to reduce damages by restricting development in flood-prone areas date back to the early 1970s. In 1975 the Centre prepared a Model Bill for the state governments to follow. Rashtriya Barh Aayog too, recommended enactment and implementation of flood plain zoning. But till this date, except for a couple of states with rarely any flood, no other state implemented it. Though it is fashionable to depict this as lack of seriousness on the part of the states, the truth probably is lack of enthusiasm for an impractical suggestion. This is a country with a total flood-prone area estimated at 45.36 mha or about 13 per cent of total geographical area. On an average 7.55 mha is affected annually by floods, of which 3.55 mha is cropped area. The area in question is not uniformly distributed between the states. More than a fourth part of the geographical area of several states is flood-prone. Must they notify, through flood zoning regulations, such vast areas as out of bounds for developmental activities? Since this is not a sensible option, no state has initiated even the flood

3. *The Hindu*, November 18, 2005.
4. *The Times of India*, 17 June 2008.

hazards mapping. Alternatively, the states may use some highly demanding standard and show only some limited parts as vulnerable zone. But if floods occur repeatedly over vast regions beyond these small patches then zoning exercise of this nature would serve no purpose. In recent years such a flood hazard map has been released by the Central government as a part of the 'vulnerability atlas' of India. A look at the map would reveal that vast areas outside the marked flood hazard zone havc witnessed repeated floods even in the last few years.

Disaster management strategies have made great strides after the UN declared the International Decade for Natural Disaster Reduction (1990-2000). Flood management is one of its components. Some progress has been made in matters like early warning of flood. But progress seems to be much less than whatever has been attained for some other disasters, like cyclone or tsunami. However, this is too early to make an assessment. In other spheres of vulnerability reduction initiatives related to flood management are missing. After the Latur earthquake, earthquake resistant construction technology has been developed. New building standards in seismic zones were introduced after the Gujarat earthquake. After Orissa Super Cyclone, a whole network of cyclone shelters were constructed, emergency evacuation plans introduced and livelihood restoration was integrated in poverty alleviation programme. After the East coast Tsunami, an excellent 'Tsunami Early Warning' system has been established. No parallel laudable initiative for flood management can be cited after any one of the the devastating floods that occur every year.

What can be Done?

Disaster management has two components: hazard control and vulnerability reduction. For flood management, both of these tasks need to be carried out simultaneously. Though inadequate, India has some programmes in both of these areas. But these are carried out by different departments, and there is only nominal coordination. Besides, the existing programmes in each of the component areas are not guided by a well-knit policy. For example, flood hazard can be reduced by suitable structural measures. Vulnerability reduction measures in such moderately flooded areas may be effective. Independently, structural measures do not eliminate flood, and vulnerability reduction has little impact in heavy flood conditions.

India has not made any attempt for a clear assessment of the available structural options. The limitations of dams and embankments in flood control are not even acknowledged. Without that their effectiveness for flood moderation cannot be assessed. Almost a century of intense technical debates following the construction of Damodar embankments in 1850s, had decided against the structure. In 1938, the National Planning Committee categorically rejected 'indiscriminate construction of flood embankments'. But the structural option made a backdoor entry in 1950s after dam construction for flood control ceased. The critical assessments that had once led Sir William Wilcocks to call those 'satanic chains' have been forgotten. Nor are there any assessments of the effectiveness of the engineering developments that have been made during the period following reintroduction in the post-Independence period.

Unlike in US or elsewhere, the responsible departments did not make any open admission about the ineffectiveness of dams and reservoirs for flood control. This, on the one hand, left room for people to view construction of new dams as solutions to flood. The flood affected population felt lack of response after nothing was done to check flood. Since there was a tacit moratorium on construction of dams primarily for flood control, the embankment construction made a semblance of responsive action. Thus an option already rejected was revived in order to appease the people. Dams do not eliminate, but possibly moderates flood. A clear admission of the limitations and possible success would have opened up room for objective assessments and systematic planning. Indeed, flood moderation is of considerable help, particularly after the developments of vulnerability reduction measures. From the experiences of several decades it is now clear that the structural measures known are not capable of eliminating flood, though they may reduce their intensities. Earlier, it was a hard fact to admit. But after categorising floods as natural disasters, along with earthquakes and cyclones, it does not seem so difficult to accept.

In this connection it is worth recollecting alternative structural measures suggested in the past. The 100 years debate of engineers in colonial India had not only rejected embankments but had also succeeded in finding possible structural alternatives. Reservoirs were one. But construction engineering was not yet so well-developed as to construct large dams and reservoirs. Another alternative, based on drainage and so-called 'overflow irrigation' was feasible. In Bengal in 1925-26, a suggestion was made by an engineer, C.A. Bentley, for using the flood flows to increase soil fertility and malaria control. This was supported by Sir William Wilcocks, a renowned water technologist of that era (Wilcocks, 1930). In 1939, a scheme, called the Damodar Hooghly Flushing and Irrigation Scheme, was passed by the Bengal Legislative Assembly. Because of political events, the construction works could not start for years. By the time the nationalist forces returned to administration technological advancements were adequate for construction of large dams. It was natural to replace the scheme by the DVC plan. But now, after we have ceased using the dam construction technology for flood control, the untested design may be worth exploring. One must not forget that it was developed by some of the best engineers of that period.

In the matter of vulnerability reduction some progress has been made in areas like early warning of flood. But effort is hesitant, in comparison to that for the disasters like cyclone or tsunami. The social costs of floods, which are far more than the actual damages, can be greatly reduced by rapid rehabilitation. Flood insurance is not the only financial instrument available for this. The others are grants or subsidies and loans. But each type of instrument has its own problem. Grants are inadequate and uncertain. The agencies distributing the grants face delivery problems. Corruption is inevitable. Loans are recoverable. But distribution face amortisation problem. Insurance seems better than both loans and grants. But the experiences of the developed countries show that these are not without problems. Private insurers are often unwilling and the government has to step in. For flood insurance, coverage is low, making premiums high. There is the usual moral hazard problem of insurance. Any attempt to implement insurance schemes in India would face two other problems—the absence of a database, which is essential for designing an insurance scheme, and the absence of an investment guideline for keeping the losses within limits. An alternative to flood insurance is weather index insurance (Barnett and Mahul, 2007), if only we knew how to predict flood from weather conditions. On a limited scale, weather index insurance for drought has been introduced in India in 2003.

Flood is one of the worst miseries suffered by India. While major earthquakes, tsunamis and other disasters occur once-in-a-while, flood damages of comparable magnitude happen every year. Historically, the unparalleled threat was well acknowledged. Flood control was high up in the development agenda of pre- and immediate post-independence India. Policies were designed, investments made, and expectations ran high. But today, though the government makes tall claims of extending protection to a large part of the flood-prone areas, available evidence lends no support to such claims. If anything, flood damages seem to be increasing. Policies and measures tried have not made any dent. Is there anything worthwhile left to be done? I am of the opinion that there is. Here I have indicated a few possibilities that need to considered and knit into a comprehensive policy. This is not an exhaustive list, but shows that possibilities have not been fully used. Considering the seriousness of the problem, comprehensive flood management is a task that should be taken up urgently.

References

Barnett, Barry J. and Olivier Mahul (2007). "Weather Index Insurance for Agriculture and Rural Areas in Lower-Income Countries", *American Journal of Agriculture Economics* 89 (5): 1241–1247.

Government of India (2006a). *Report of the Working Group on Water Resources for the XIth Five Year Plan (2007-12)*. New Delhi: Ministry of Water Resources. p.126.

———. (2006b). *Ninth Five Year Plan*. Volume II. New Delhi: Planning Commission.

———. (2008). "Agriculture, Rural Development, Industry, Services and Physical Infrastructure", *Eleven Five Year Plan (2007-2012)*.Volume III. New Delhi: Planning Commission. p.52.

Lin, T., Franklin D. De Guzman and Maria Cita Cuevas (2007). "Flood Insurance as a Flood Management Tool: An Economic Perspective", *ERD Working Paper* No. 99. Manila: ADB.

Ministry of Home Affairs (n.d.). *Hazards, Disasters and Your Community*, Version 1.0 National Disaster Management Division, GoI.

Planning Commission (2002). "Irrigation, Flood Control and Command Area Development", *Tenth Five Year Plan 2002-07*, Part II, Chapter 8.1. GoI.

———. (2007). *Report of The Steering Committee on Water Resources For Eleventh Five Year Plan (2007-2012)*.

Swiss Re (2007). "Natural Catastrophes and Man-made Disasters in 2006—Low Insured Losses", *Swiss Re Sigma* No. 2/2007. Zurich: Swiss Reinsurance Co.

The Hindu (2005). "Jayalalithaa Seeks Rs. 1,742 cr. for Flood Relief from Calamity Fund", *The Hindu*, Friday, Nov 18.

The Times of India (2008). June 17. *http://articles.timesofindia.indiatimes.com/2008-06-17/patna/27770539_1_chief-minister-nitish-kumar-healthy-centre-state-relationship-calamity-relief-fund*

Wilcocks, William (1930). *Lectures on the Ancient System of Irrigation in Bengal and its Application to Modern Problems*. Calcutta: University of Calcutta.

11

Climate Sensitivity of Indian Agriculture

Role of Technological Development and Information Diffusion

K.S. Kavi Kumar

Over the past two decades the debate on global climate change has moved from scientific circles to policy circles with the world nations more seriously than ever exploring a range of response strategies to deal with this complex phenomenon that is threatening to have significant and far reaching impacts on human society. The Intergovernmental Panel on Climate Change (IPCC) in its fourth assessment report observed that, 'warming of climate system is now unequivocal, as is now evident from observations of increases in global average air and ocean temperatures, widespread melting of snow and ice, and rising global sea level' (Solomon *et al.*, 2007). Policy responses to climate change include mitigation of greenhouse gases (GHGs) that contribute to the expected changes in the Earth's climate, and adaptation to potential impacts caused by the changing climate. While GHG mitigation policies have dominated the overall climate policy so far, adaptation strategies are also being emphasised now to form a more comprehensive overall policy response.

The United Nations Framework Convention on Climate Change (UNFCCC)—the international apex body on climate change—refers to adaptation in the context of change in climate only. In other words without GHG emissions there is no climate change and hence no need for adaptation. Going by this widely accepted interpretation, adaptation is necessary only because mitigation of GHGs may not completely halt climate change. Stern Review summarises this view: 'adaptation is crucial to deal with the *unavoidable* impacts of climate change to which the world is already committed' (Stern, 2006: emphasis added). However, the reality is that several millions of people world over are currently at the risk of climate related impacts. To say that their sufferings are not the concern of climate change policy could make such policies irrelevant from the point of view of developing countries where most victims of climate related impacts reside. At the same time, it may not be meaningful to attribute every climate related issue to the climate change policy because it would make an already complex policy issue further complicated. Adaptation in the context of climate change should be discussed taking into account these conflicting perspectives.

For both mitigation and adaptation policy formulation, one of the crucial inputs needed is the potential impacts due to climate change on various climate sensitive sectors. For mitigation, such information would provide the required justification for decarbonising the energy systems. On the other hand, in the context of adaptation, knowledge on climate change induced impacts will be helpful in prioritising the adaptation in the most needed sectors and regions. Further, climate change impacts estimated with proper accounting of adaptation will be helpful in identifying the factors that ameliorate the adverse effects of climate change.

Concern about potential adverse impacts of climate change has triggered what could be named as impact assessment research since early 1990s. Over the past two decades the impact assessment research has gone through a metamorphosis with gradual movement towards vulnerability and adaptation assessment. However, impact assessment studies continue to prosper with interesting innovations in methodology and scope of analysis. Figure 11.1 shows the conceptual basis for climate change impact assessment studies, with focus on agriculture. The reference case illustrates the future without climate change. The naïve farmer allows the adverse impacts of the climate change to manifest without undertaking any adaptation. The impacts under such scenario are referred as potential impacts. A typical farmer is expected to undertake autonomous (or, unplanned) adaptation and hence avoid impacts to some extent (resultant impacts are referred as expected impacts). With planned (along with autonomous) adaptation, a smart farmer can further reduce the climate change impacts; the resultant impacts referred as residual impacts. A clairvoyant farmer on the other hand, in principle, can undertake all theoretically possible adaptations and avoid climate change impacts to the maximum extent. The impacts that even the clairvoyant farmer can not avoid are referred as unavoidable impacts of climate change. In the climate change impacts literature the residual impacts (or in some cases, the unavoidable impacts) are referred as 'vulnerability'.

Figure 11.1

Conceptualisation of Climate Change Impacts

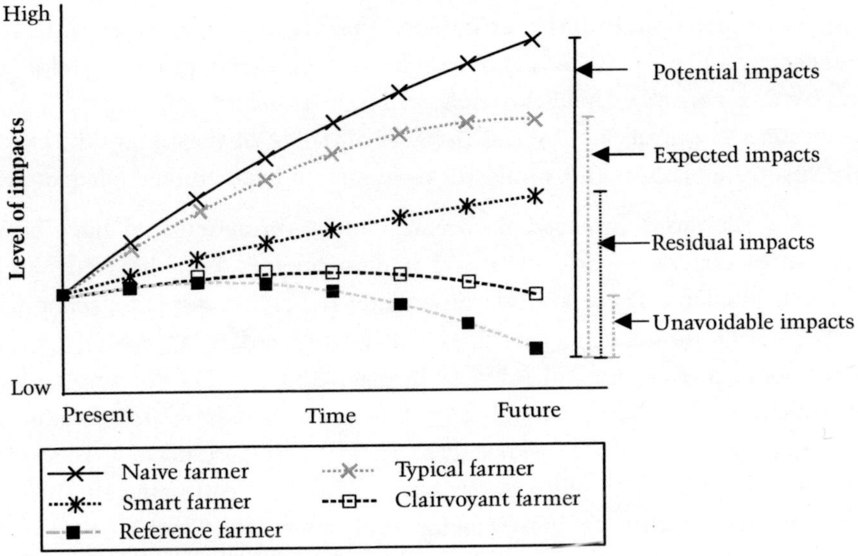

Source: Fussel and Klein (2006).

Climate Change and Indian Agriculture

India is a vast country covering 3.28 million km^2, occupying only 2.4 per cent of the world's geographical area but supporting 16.2 per cent of the global human population. It is endowed with varied climate supporting rich biodiversity and highly diverse ecology. More than 60 per cent of its population is dependent on climate sensitive activities such as agriculture. Climate change projections made up to year 2100 for India, indicate an overall increase in temperature by 2-4°C with no substantial change in precipitation quantity. The expected changes in climate, especially rainfall, are also marked by significant regional variation, with the Western Ghats, the Central Indian and the Northeastern regions projected to receive more rainfall compared to the other parts of India. Further, an increase in intensity and frequency of extreme events such as droughts, floods and cyclones is also projected. All these changes are likely to have adverse impacts on India's water resources, agriculture, forests and other ecosystems, coastal zones, energy and infrastructure and on human health. Agricultural impacts due to climate change have received considerable attention in India as they are closely linked to the food security and poverty status of a vast majority of population.

Mall *et al.* (2006) provides an excellent review of climate change impact studies on Indian agriculture mainly from physical impacts perspective. The available evidence shows significant drop in yields of important cereal crops like rice and wheat under climate change conditions. However, biophysical impacts on some of the important crops like sugarcane, cotton and sunflower have not yet been studied adequately.

The economic impacts of climate change on agriculture have been studied extensively world over and it continues to be a hotly debated research problem. There are two broad approaches for assessing economic impacts—agronomic-economic approach and Ricardian approach. In the first approach, the physical impacts (in the form of yield changes and/or area changes estimated through crop simulation models) are introduced into an economic model exogenously as Hicks neutral technical changes. In the Indian context, Kumar and Parikh (2001a) have estimated the macro level impacts of climate change using such approach. They showed that under doubled carbon dioxide concentration levels in the later half of 21st century, the gross domestic product would decline by 1.4 to 3 percentage points under various climate change scenarios. More significantly they also estimated increase in the proportion of population in the bottom income groups of the society in both rural and urban India under climate change conditions. One of the major limitations of this approach is its treatment of adaptation. Since the physical impacts of agriculture are to be re-estimated under each adaptation strategy, only a limited number of strategies can be analysed.

Since the scope for incorporating adaptation into the agronomic-economic approach is rather limited, an alternative approach was proposed by Mendelsohn *et al.* (1994). This approach, called Ricardian approach, is similar to Hedonic pricing approach of environmental valuation. The approach is based on the argument that, 'by examining two agricultural areas that are similar in all respects except that one has a climate on average (say) 3°C warmer than the other, one would be able to infer the *willingness to pay* in agriculture to avoid a 3°C temperature rise' (Kolstad, 2000). While all possible adaptations are accounted for in the impact estimation using this approach, the constant relative prices assumption could lead to biases in the results of this approach. For India, Kumar and Parikh (2001b) have used a variant of this approach and showed that a 2°C temperature rise and 7 per cent increase in rainfall would lead to almost 9 per cent loss in farm level net revenue (1990 net revenue expressed in 1999-2000 prices). The regional differences are significantly large with Northern and Central Indian districts along with coastal districts bearing relatively large impact. Mendelsohn *et al.* (2001) have compared climate sensitivity of the US, Brazilian and Indian agriculture using the estimates based on the Ricardian approach and have argued that using the US estimates for assessing climate change impacts on Indian agriculture would provide biased results.

The results of the two broad approaches outlined above correspond to what could be termed as 'typical' and 'clairvoyant' farmer, respectively (see Figure 11.1). While the estimates from agronomic-economic approach account for adaptation only in partial manner, the Ricardian approach treats farmer as though s/he has perfect foresight. In the Ricardian approach farmers are assumed to identify instantaneously and perfectly any change in climate, evaluate all associated changes in market conditions and then modify their actions to maximise profits. These assumptions also imply that agricultural system is ergodic—i.e., space and time are substitutable. Ergodic assumption imply, for example, that skills, institutional and financial endowments for responding to say, drought (that are typically refined in arid places) are assumed to be available for use by people in humid areas (where such resources are underdeveloped) immediately and in essentially costless manner. Further there is scope for inter-farmer communication and information diffusion. Both these factors motivate incorporation of spatial features in the Ricaridan analysis. There are other motivations for accounting for spatial autocorrelation in the Ricardian analysis. For instance, Schlenker *et al.* (2005) bring in spatial features to arrive at efficient estimators of regression parameters. Recent studies in the US have demonstrated that such refinements are essential to get accurate estimates of climate sensitivity (Polsky, 2004; Schlenker *et al.*, 2006).

Similarly, careful analysis of the changing nature of climate sensitivity of Indian agriculture is important to understand the role of technology in ameliorating the climate change impacts. This paper attempts to incorporate these features into the Ricardian approach to assess the climate change impacts on Indian agriculture.

Objective of the Paper

Any study dealing with climate change impact assessment should be wary about the potential uncertainties associated with the analysis. Uncertainties exist at each step of the impact assessment starting with the climate predictions. The degree of uncertainty increases as the analysis moves towards economic impact assessment. Among the various sources of uncertainties, the uncertainty associated with climate predictions perhaps is the most crucial and difficult to deal with. Most studies in the literature have adopted scenario approach to handle the climate uncertainty. Thus, there is no single climate prediction for which the impacts would be assessed, but they are estimated for a range of likely temperature/precipitation combinations.

The present study also follows similar approach and as explained at the beginning of this Section III, predictions from a number of climate models have been considered while developing India-specific climate change scenario. Based on this suite of climate models, by 2070-2099, the average temperatures across seasons are likely to range between 2.5 to 5°C over different regions of India. Winter temperatures are likely to be significantly higher ranging between 3.75 to 4.95°C across the regions. The uncertainty is much higher with regard to the precipitation prediction and by 2070-2099 the Southwest monsoon precipitation is likely to increase between 9 to 27 per cent across the regions.

The rest of the paper is structured as follows: Section IV explains the model structure and data. Section V presents results and discusses the distributional issues of climate change impacts on Indian agriculture. Section VI briefly discusses the lessons learned about inter-farmer communication through focus group meetings in Andhra Pradesh and Tamil Nadu. Finally, the last Section provides policy discussion and concludes the paper.

Model Specification and Data

The Ricardian approach suggests that under certain assumptions, the value of the change in the environment is captured exactly by the change in aggregate land values (see Mendelsohn *et al.*, 1994 for more details). Due to non-existent and/or absence of well functioning land markets in the developing countries, a variant of Ricardian approach has been used in the earlier Indian studies. In place of land values, farm level net revenue is used as welfare indicator and the value of the change in the environment is assessed through change in farm level net revenue. The Ricardian model is specified as follows:

$$R = f(T_j, T_j^2, P_j, P_j^2, T_jP_j, SOIL, BULLOCK, TRACTOR, POPDEN, LITPROP, CULTIV, HYV, IRR, ALT) \quad (1)$$

where, R represents farm level net revenue per hectare in constant rupees; T and P represent temperature and rainfall respectively. It may be noted that based on the existing literature a quadratic functional specification is adopted and the model includes climate interaction terms also. The control variables include soil (captured through dummies representing several soil texture classes and top-soil depth classes), extent of mechanisation (captured through number of bullocks and tractors per hectare), percentage of literate population, population density, altitude (to account for solar radiation received), number of cultivators (since the cost of labour could not be accounted for while calculating the dependent variable), fraction of area under irrigation and fraction of area under high yielding variety (HYV) seeds.

Cross-sectional data is used for estimating the above model. Districts are the lowest administrative unit at which reliable agricultural data is available. A comprehensive district level dataset of the period 1956 to 1999 is developed for the purpose of analysis. Agricultural data at district level is assembled in the dataset along with relevant demographic and macro-economic data. This dataset expands an earlier dataset developed by the author along with two other researchers for the period 1956 to 1986. The dataset covers districts defined as per 1961 Census across 13 major states of India (Andhra Pradesh, Haryana, Madhya Pradesh, Maharashtra, Karnataka, Punjab, Tamil Nadu, Uttar Pradesh, Bihar, Gujarat, Rajasthan, Orissa and West Bengal). In all, the dataset includes 271 districts. For the purpose of analysis, the dataset is divided into three distinct periods: 1956-1970; 1971-1985; 1986-1999. Analysis over these three periods will provide insight on changing nature of climate sensitivity of Indian agriculture over time.

The variables covered in the dataset include, gross and net cropped area; gross and net irrigated area; cultivators; agricultural labourers; cropped area under HYV seeds; total cropped area under five major crops (rice, wheat, maize, *bajra* and *jowar*) and 15 minor crops (barley, gram, *ragi*, *tur*, potato, groundnut, tobacco, sesamum, ramseed, sugarcane, cotton, other pulses, jute, soyabean and sunflower); bullocks; tractors; literacy rate; population density; fertiliser consumption (N, P, K) and prices; agricultural wages; crop produce; farm harvest prices; soil texture and top soil depth. For the purpose of analysis, farm level net revenue per hectare is defined as follows:

$$\text{Net revenue per ha} = \frac{((\text{gross revenue}) - (\text{fertiliser and labour costs}))}{\text{Total area}}$$

where, gross revenue is calculated over 20 crops mentioned above, total area is the cropped area under the 20 crops, fertiliser costs are total yearly costs incurred towards use of fertiliser for all the crops and labour costs are yearly expenses towards hiring agricultural labourers. It may be noted that costs attributable to cultivators, irrigation, bullocks and tractors are not included in the net revenue calculations as appropriate prices are difficult to identify. However, these variables are used as control variables in the model as specified in Equation (1).

Unfortunately, there is no 'clean' climate data available for the analysis. Meteorological data is typically collected at meteorological stations and any district may have one or many stations with in its boundary. Since all other data is attributable to a hypothetical centre of the district, the climate data should also be worked out at the centre of the district. For this purpose meteorological station data is interpolated to arrive at district-specific climate (see, Kumar and Parikh, 2001b; Kumar, 2003 for more details on the surface interpolation employed for generate district level climate data). Climate data corresponding to about 391 meteorological stations spread across India is used for the purpose of developing district level climate. The data on climate—at the meteorological stations and hence at the districts—correspond to the period 1951-1980 and is sourced from a recent publication of India Meteorological Department. All the climate variables are represented through four months—January, April, July and October. The climate variables include daily mean temperature and monthly total rainfall.

In each case the panel data is analysed with fixed year effects—it may be noted that district effects are not considered as the climate (and soil) data is invariant over time. Fixed and random year effects specification is tested through Hausman test in each case. In each time period, fixed effects specification could not be rejected in favour of random effects model. Further, since the units of analysis (i.e., districts) differ significantly in size and agricultural activities, the measurement errors might also substantially differ across districts. Hence, the data for each unit of analysis is weighted by the total area under the 20 crops.

Climate Sensitivity and Spatial Autocorrelation

As argued in the first section, presence of spatial autocorrelation necessitates re-specification of model either as spatial lag or spatial error model as shown below:

Spatial error model: $y = X\beta + \eta, \text{where } \eta = \rho W\eta + \varepsilon$ (2a)

Spatial lag model: $y = \rho Wy + X\beta + \varepsilon$ (2b)

where, y is (nx1) vector of dependent variable observations, X is (nxm) matrix of observations on independent variables, β is (mx1) regression coefficient vector, η is (nx1) vector of spatially correlated error terms, ρ is (1x1) the spatial autoregressive parameter, W is (nxn) spatial weights matrix, ε is (nx1) vector of random error terms. Note that y and X are respectively, the left hand and right hand side variables specified in Equation (1) above. The period 1966-1986 is considered for the spatial analysis.

One of the crucial inputs needed for spatial analysis is the weight matrix W. While there are several ways to generate the weight matrix, the present analysis used rook contiguity based weight matrix generated for the Indian districts in GeoDa software.[1] Since it is not feasible to estimate the

1. Spatial econometric software developed by Prof Luc Anselin of University of Illinois (version 0.9.5).

spatial fixed effects model in GeoDa, the weight matrix is transferred via R-software to American Standard Code for Information Interchange (ASCII) data format. The spatial panel model is estimated using MATLAB software as it provides scope for reading sparse matrices. Table 11.1 shows details of various analyses carried out.

Table 11.1

Details of Various Analyses

	Aim of the Analysis	*Period(s) of Analysis*	*Model Specification*	*Estimation Procedure and Software Used*
1.	Explore changing nature of climate sensitivity over time	1956-1999 with sub-periods: 1956-1970; 1971-1985; 1986-1999	Equation (1)	Panel fixed (year) effects by weighting the observations; STATA
2.	Explore influence of spatial autocorrelation on climate sensitivity	1966-1986	Equation (2a) and (2b)	Panel fixed (year) effects with correction for spatial autocorrelation; GeoDA; R; MATLAB

Results

The results are reported in three subsections: in the first subsection the changing nature of climate response function over time is presented along with estimates of climate change impacts. The second subsection reports the results based on spatial analysis along with the estimates of climate change impacts with and without the correction for spatial autocorrelation.

The climate change projections for India used for the analysis are those reported in Cline (2007). The climate change projections are average of predictions of six general circulation models including HadCM3, CSIRO-Mk2, CGCM2, GFDL-R30, CCSR/NIES and ECHAM4/OPYC3. Table 11.2 shows the region-wise and season-wise temperature and rainfall changes for the period 2070-2099 with reference to the base period 1960-1990. From these regional projections, state-wise climate change predictions are assessed by comparing the latitude-longitude ranges of the regions with those of the states. Besides this, India-specific climate change scenario, the impacts are also assessed for two illustrative uniform climate change scenarios that embrace the aggregate changes outlined in the fourth assessment report of IPCC (Solomon *et al.*, 2007).

Table 11.2

Projected Changes in Climate in India: 2070-2099

	Jan.-March	*April-June*	*July-Sep.*	*Oct.-Dec.*
Temperature change (°C)				
Northeast	4.95	4.11	2.88	4.05
Northwest	4.53	4.25	2.96	4.16
Southeast	4.16	3.21	2.53	3.29
Southwest	3.74	3.07	2.52	3.04
Precipitation change (%)				
Northeast	-9.3	20.3	21.0	7.5
Northwest	7.2	7.1	27.2	57.0
Southeast	-32.9	29.7	10.9	0.7
Southwest	22.3	32.3	8.8	8.5

Changing Nature of Climate Sensitivity of Indian Agriculture Over Time

Equation (1) is estimated using the pooled data over the period 1956-1999 by separating out climate coefficients for three distinct periods: 1956-1970, 1971-1985 and 1986-1999. Year effects are included in the estimation. Hausman test favoured fixed effects specification against the random effects. Table 11.3 shows the estimates of climate coefficients along with important control variables for the three time periods. The dependent variable in each case is net revenue per hectare expressed in constant 1999-2000 prices. The control variables are all significant in all the three periods and have expected sign. Barring a very few exceptions, in all the three periods the climate coefficients are all significant and the F-tests for joint significance of climate coefficients in each period rejected the null-hypothesis.

Inclusion of interaction terms makes it difficult to interpret the marginal effects of temperature and precipitation. To gain insight about the effect of climate variation terms in the model, the climate change induced impacts are estimated for the climate change scenarios described above.

The climate change induced impacts are measured through changes in net revenue triggered by expected changes in the climate variables. The impacts are estimated for each year at individual district level and are then aggregated to derive the national level impacts. Average impacts over all the

Table 11.3

Climate Response Function Over Time

	1956-1970		*1971-1985*		*1986-1999*	
Variable	*Coefficient*	*p-value*	*Coefficient*	*p-value*	*Coefficient*	*p-value*
Climate variables						
January-T	-449.8602	0.000	-327.4542	0.001	-399.9256	0.000
April-T	-26.2412	0.809	-855.2145	0.000	-985.8106	0.000
July-T	-737.5402	0.000	-838.6752	0.000	-763.3346	0.000
October-T	1603.5600	0.000	2158.3240	0.000	2624.1480	0.000
January-P	17.0548	0.189	39.9249	0.001	122.5294	0.000
April-P	-8.0753	0.038	-19.5432	0.000	-16.7260	0.000
July-P	-0.2700	0.755	-2.9057	0.000	1.0216	0.194
October-P	25.8997	0.000	26.7250	0.000	12.4678	0.002
January-T-sq	-6.2100	0.702	-49.9343	0.001	26.6250	0.111
April-T-sq	-15.1834	0.605	150.2279	0.000	50.1199	0.049
July-T-sq	-157.3086	0.007	-88.3555	0.109	-350.6405	0.000
October-T-sq	-154.4181	0.000	-269.8104	0.000	-321.1355	0.000
January-P-sq	-0.7280	0.069	-2.9675	0.000	-3.0880	0.000
April-P-sq	0.1024	0.003	0.1896	0.000	0.2141	0.000
July-P-sq	0.0039	0.016	0.0016	0.276	0.0030	0.034
October-P-sq	-0.0142	0.686	0.0459	0.161	-0.0654	0.049
January-TP	-21.7075	0.000	-34.5983	0.000	-20.2050	0.000
April-TP	7.9786	0.000	16.5937	0.000	15.8305	0.000
July-TP	-1.2940	0.022	-1.8498	0.000	-2.0714	0.000
October-TP	1.1694	0.546	-3.1449	0.074	-6.1247	0.001
Control variables						
Cultivators/ha	336.7072	0.263	435.8161	0.068	587.3252	0.009
Bullocks/ha	958.3203	0.009	-200.0078	0.484	-727.1433	0.006
Tractors/ha	676432.5000	0.000	152806.9000	0.000	88268.4800	0.000
Literacy	123.9629	0.873	2829.1690	0.000	3326.1930	0.000
Pop. density	376.6634	0.000	217.0976	0.000	47.3803	0.019
Irrigation %	4442.7910	0.000	2178.5120	0.000	2091.4530	0.000
No. of obs.			11924			
Adj R^2			0.5398			

years are reported in Table 11.4. The table reports the all-India level impacts estimated in each time period as percentage of 1990 all-India net revenue expressed in 1999-2000 prices. Comparison with 1990 net revenue is considered mainly to accommodate comparison with previous results reported in the literature. The impacts are interpreted as change in 1990 net revenue if the future climate changes were to be imposed on 1990 economy. As could be seen the impacts (based on the illustrative uniform scenarios) are increasing over time indicating increasing climate sensitivity of Indian agriculture. This is despite the possible advances made through technology adoption and overall development. Significantly higher impacts reported in the period from mid-1980s to late 1990s. This finding corroborates the growing evidence of weakening agricultural productivity over the similar period in India. The impacts estimated using India-specific climate projections show that impacts declined in the period 1971-1985 and again increased in the last period. The decline in the middle period could possibly be due to improved resilience of Indian agriculture during this period and also due to the regional variation in the climate projections.

Table 11.4

Climate Change Impacts Over Time

Scenario	*1956-1970*		*1971-1985*		*1986-1999*	
	Impacts	*% of 1990 Net Revenue*	*Impacts*	*% of 1990 Net Revenue*	*Impacts*	*% of 1990 Net Revenue*
+2°C/7%	-53.7	-6.1	-76.8	-8.7	-188.7	-21.3
+3.5°C/14%	-297.4	-33.6	-303.4	-34.3	-754.9	-85.3
India-specific CC scenario	-219.6	-24.8	-153.6	-17.4	-544.4	-61.5

Note: Impacts are in billion rupees, 1999-2000 prices; 1990 net revenue is also expressed in 1999-2000 prices.

Effect of Spatial Autocorrelation on Climate Sensitivity

The spatial clustering of the dependent variable (i.e., net revenue per hectare) is analysed by constructing Moran scatter plots for several time points in the period 1956-1999. Figure 11.2 shows the scatter plots along with the Moran's I value. The scatter plot is graph of *Wy versus y*, where *W* is a row-standardised spatial weight matrix and *y* = [(net revenue – mean net revenue)/standard deviation of net revenue]. Clustering of values in the upper right quadrant and lower left quadrant represents significant positive spatial autocorrelation. As could be seen from Figure 11.2 in all the six periods for which the scatter plots are reported, the dependent variable exhibited significant positive spatial autocorrelation.

Indication of significant pattern of spatial clustering given by the spatial autocorrelation statistic represents only the first step in the analysis of spatial data. Two typically considered specifications for modelling spatial dependence are: spatial error and spatial lag model. These models specified in Equations (2a) and (2b) are estimated for the period 1966-1986. Table 11.5 shows the climate response functions estimated with and without consideration of spatial autocorrelation. All the estimates are based on fixed (year) effects specification in the pooled data and observations are weighted by the total area under all the crops considered in the analysis. Barring a few exceptions, the climate coefficients in the models that accounts for spatial autocorrelation (either through spatial lag or spatial error models) are uniformly lower than that ignores the presence of spatial autocorrelation indicating the true climate change impacts to be lower. This is confirmed by the climate change impacts reported in Table 11.6. The overall impacts estimated (for same climate change scenario) using climate coefficients obtained from model that accounts for spatial autocorrelation are significantly lower than those obtained from model that ignores the spatial effects. Figures 11.3 and 11.4 compare the distribution of climate change impacts at the state and district level between the model accounts for spatial autocorrelation and that does not.[2] As could be observed, consideration of spatial effects has contributed to positive spin-offs, possibly among other factors due to the inter-farmer communication. The following section explores this issue further.

2. Only spatial lag model results are reported for the purpose of comparison.

Figure 11.2

Spatial Autocorrelation: Moran Scatter Plots

(a) 1960: Moran's I = 0.479

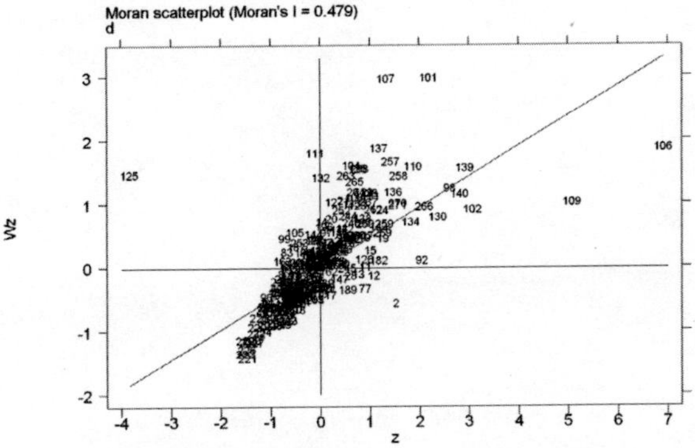

(b) 1970: Moran's I = 0.223

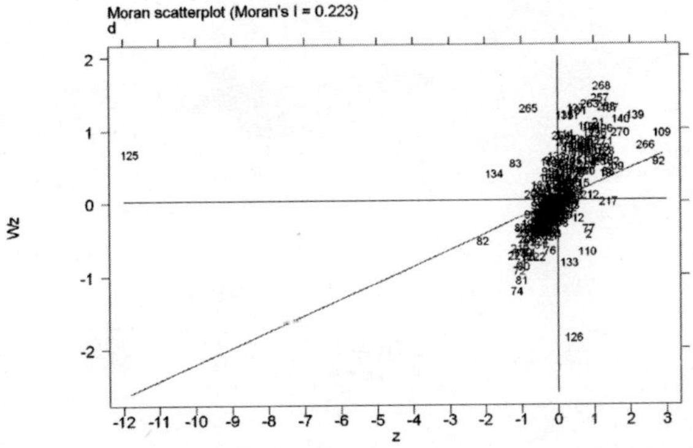

(c) 1980: Moran's I = 0.395

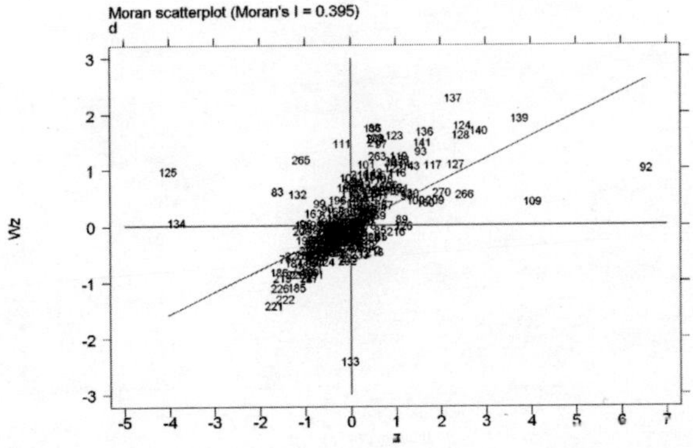

(d) 1985: Moran's I=0.431

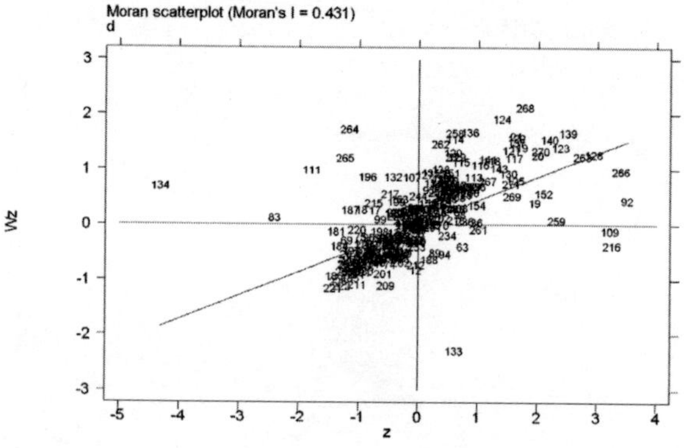

(e) 1990: Moran's I=0.209

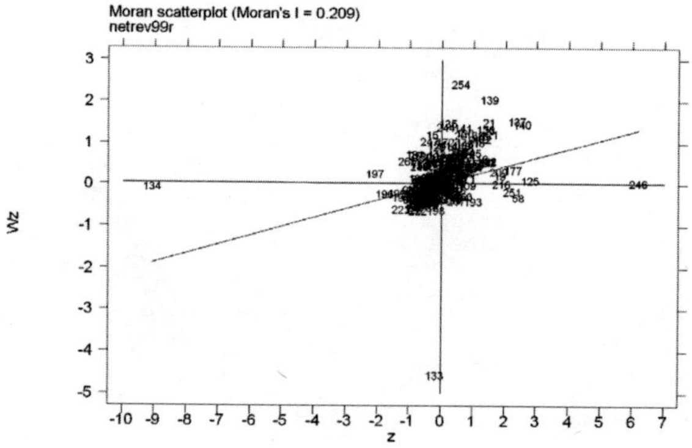

(f) 1995: Moran's I=0.207

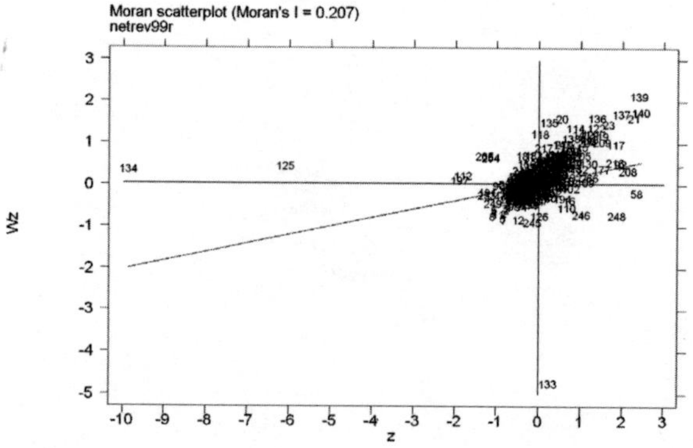

Table 11.5

Effect of Spatial Autocorrelation on Climate Sensitivity: 1966-1986

	Without Spatial Autocorrelation		With Spatial Autocorrelation			
			Spatial Lag Model		Spatial Error Model	
Variable	Coefficient	p-value	Coefficient	p-value	Coefficient	p-value
Climate variables						
January-T	-443.3306	0.000	-394.8037	0.000	-395.3314	0.000
April-T	-695.5482	0.000	-537.1341	0.000	-668.5077	0.000
July-T	-817.8726	0.000	-575.3063	0.000	-809.3126	0.000
October-T	2160.3500	0.000	1833.0273	0.000	1709.0474	0.000
January-P	38.5183	0.000	13.5505	0.106	-7.2706	0.448
April-P	-17.2423	0.000	-14.5812	0.000	-7.8016	0.004
July-P	-2.1913	0.000	-1.2732	0.027	-2.5012	0.000
October-P	29.5283	0.000	20.7544	0.000	18.3900	0.000
January-T-sq	-43.8116	0.000	-24.1114	0.033	-11.4262	0.332
April-T-sq	118.3944	0.000	101.9399	0.000	138.9808	0.000
July-T-sq	-96.9162	0.014	-25.5708	0.524	117.6600	0.006
October-T-sq	-264.0382	0.000	-233.9736	0.000	-236.2654	0.000
January-P-sq	-2.7809	0.000	-2.5723	0.000	-1.8918	0.000
April-P-sq	0.1738	0.000	0.1530	0.000	0.0967	0.000
July-P-sq	0.0037	0.001	0.0049	0.000	0.0022	0.039
October-P-sq	0.0281	0.232	0.0885	0.000	0.0566	0.019
January-TP	-36.3116	0.000	-38.5205	0.000	-26.7720	0.000
April-TP	15.8434	0.000	15.2391	0.000	10.3268	0.000
July-TP	-1.4774	0.000	-0.6823	0.071	-0.3687	0.346
October-TP	-2.8658	0.024	-4.1127	0.001	1.8310	0.192
Control variables						
Cultivators/ha	253.3993	0.119	163.0752	0.331	758.4805	0.000
Bullocks/ha	103.0263	0.615	558.4283	0.009	1105.5576	0.000
Tractors/ha	147348.6918	0.000	63282.7868	0.000	67538.9637	0.000
Literacy	2429.2926	0.000	4039.0490	0.000	3160.1950	0.000
Pop. density	178.9486	0.000	174.4978	0.000	182.0459	0.000
Irrigation %	2669.4620	0.000	2648.3836	0.000	3538.1377	0.000
Spatial lag/spat. auto.			0.0649	0.000	0.5719	0.000
No. of obs.	5691		5691		5691	
Adj. R^2	0.6791		0.6517		0.7233	

Source: Author's calculation.

Table 11.6

Climate Change Impacts: Without and With Accounting for Spatial Autocorrelation (1966-1986)

Scenario	*Without Spatial Autocorrelation*		*With Spatial Autocorrelation*			
			Spatial Lag Model		*Spatial Error Model*	
	Impacts	*% of 1990 Net Revenue*	*Impacts*	*% of 1990 Net Revenue*	*Impacts*	*% of 1990 Net Revenue*
+2°C/7%	-81.2	-9.17	14.2	1.6	-22.9	-2.6
India-specific CC scenario	-195.1	-22.1	43.4	4.9	-2.1	-0.23

Note: Impacts are in billion rupees, 1999-2000 prices; 1990 net revenue is also expressed in 1999-2000 prices.

Source: Author's calculation.

Figure 11.3

State-wise Distribution of Impacts due to India-specific Climate Change Scenario: With and Without Spatial Autocorrelation

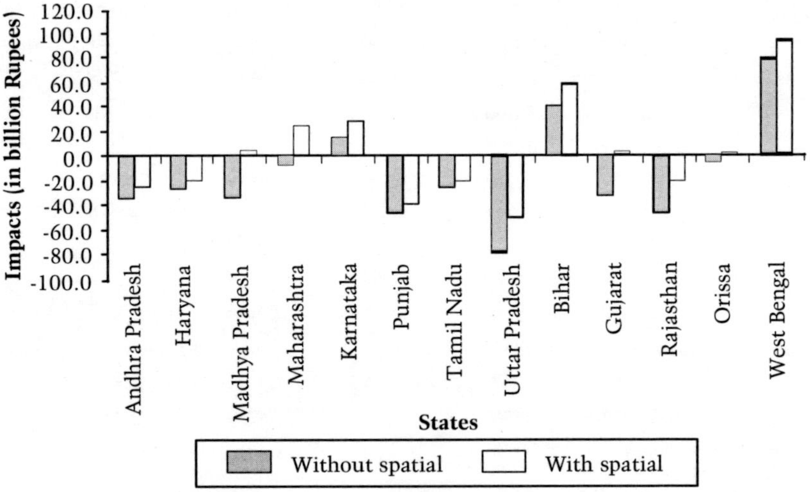

Source: Author's calculation.

Figure 11.4

Distribution of Climate Change Impacts across Districts: Without and With Spatial Autocorrelation

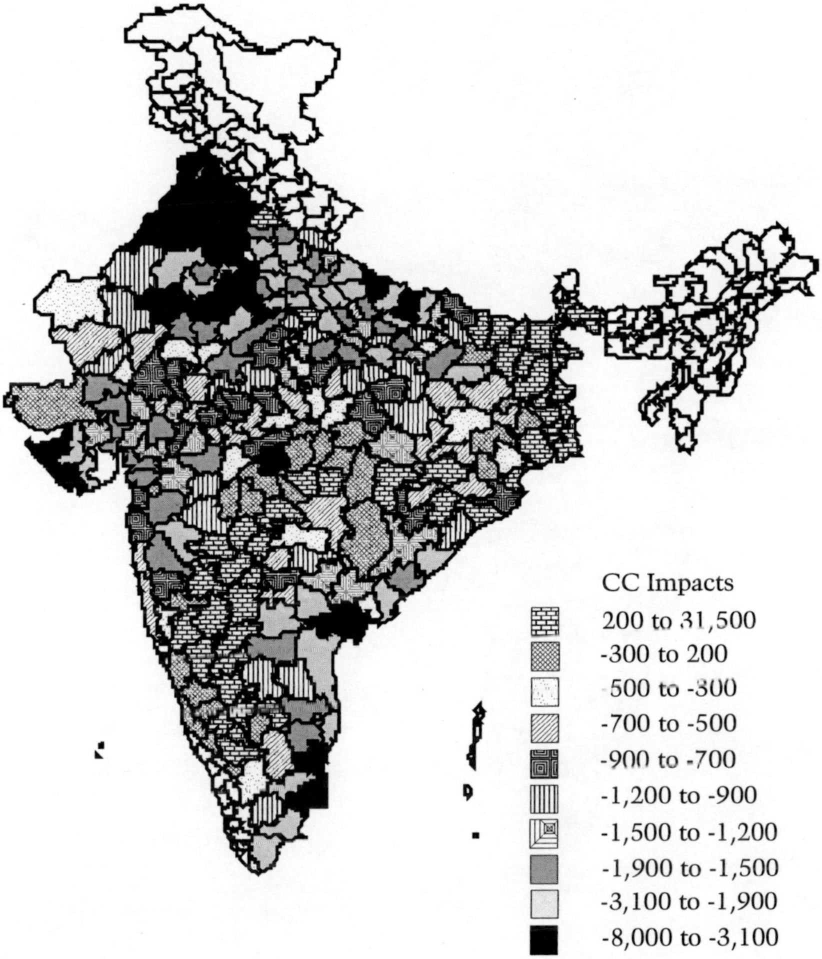

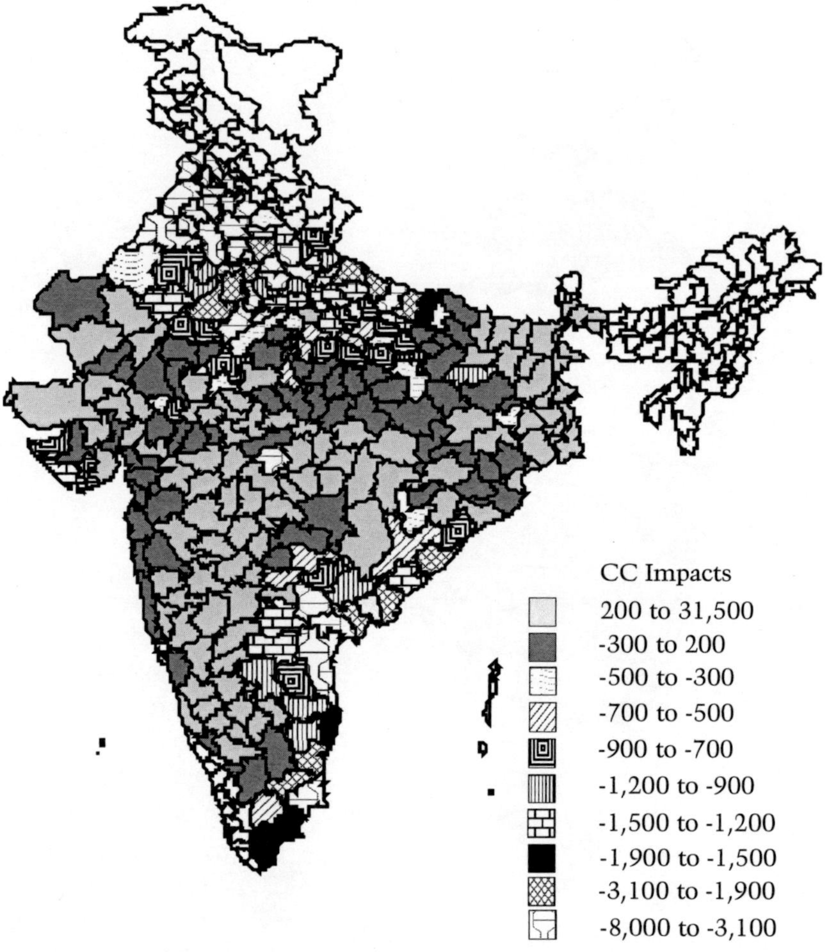
CC Impacts
200 to 31,500
-300 to 200
-500 to -300
-700 to -500
-900 to -700
-1,200 to -900
-1,500 to -1,200
-1,900 to -1,500
-3,100 to -1,900
-8,000 to -3,100

Evidence on Inter-Farmer Communication

In an attempt to understand the scope and extent of information exchange between farmers, focus group meetings are held at around six villages each in Tamil Nadu and Andhra Pradesh.[3] The focus group meetings mainly explored the perceptions of the villagers about the climate change and their views on strategies helpful in ameliorating the climate change impacts. Among other things, special attention is paid to the channels through which information diffusion takes place.

The field level analysis showed that while most farmers are familiar with the term climate change, their understanding is often completely misplaced. All climate/natural patterns are perceived as climate change with little or no distinction between future climate change and preset day climate concerns (that manifest in the form of climate extremes like droughts, floods and cyclones, and abnormal weather patterns like un-seasonal rainfall etc.). Also, there is a consensus in most discussions that anthropogenic activities leading to excess pollution are often responsible for the abnormal weather.

Most farmers also consider climate/weather concerns to be more threatening than other risks, such as price changes. The reasons cited for such perceptions include, bigger scale of impact that climate/weather risks may cause, and limited scope for adaptation. Such perceptions are uniformly held by small, medium and large farmers.

Almost all focus group meetings indicated that there is dearth of information. Farmers irrespective of size are in search of information—which could include advice on input use, pest control, agronomic practices, and soil and water conservation practices. Among the various sources through which information diffusion takes place, most focus group discussions ranked large farmers in the neighbourhood as the primary source. Not surprisingly, the agricultural extension services offered by the government are not seen as appropriate source of information, mainly due to the manner in which the extension services provide information. While the information needs are different across farmers based on their scale of

3. The field studies are carried out during the months of March-April 2008 with the help of local NGOs.

operation and kind of crops cultivated, the agricultural extension services often package the information in uniform manner as though one size fits all. Similarly, the usual information diffusion sources such as television and radio also appear to be less effective in reaching out, partly because these sources are often seen as entertainment sources rather than information channels. Discussion in several focus group meetings revealed that farmers often depend on fertiliser and pesticide dealers for information on new varieties and new agricultural practices. While this source has appropriate self-regulated checks against provision of wrong information, it is important to ensure that incorrect information does not reach the farmers even inadvertently.

New information does not often reach agricultural labourers. Given the large size of this group and the important role it plays in determining agricultural productivity, it is important to ensure that this group is also targetted along with farmers in providing information on agricultural practices etc. Similarly, the information diffusion must take place to reach female farmers also alongside their male counterparts.

The field studies also revealed that new sources of information diffusion should be explored and experimented. Given the fragmented nature of Indian agricultural lands, large scale participation of corporate sector in providing agricultural extension services could be difficult, and hence other options must be explored. Among other things, the farmers favoured participation of agricultural cooperatives, NGOs and dealers of inputs and fertilisers in information diffusion. In this context, other country experiences should also be carefully studied to identify the routes through which the agricultural extension services could be provided to the farmers. For instance, in Ecuador the agricultural extension workers operate in tandem with the farmers through share cropping to ensure proper information diffusion. On the other hand, Chile finances the costs of private sector firms transferring the technology know-how and information on new agricultural practices to small scale farmers.

Conclusions

Before discussing the policy issues emerging from the study, it is important to place the climate change issues in the broader context of sustainable development. Given the long-time horizon nature of climate change problem, the climate change operates at the margin along with several other changes over the next several decades. Thus, climate change induced threat operates at the 'margin' of significant threat that societies (especially in developing countries) face through other factors. For instance, quoting other studies Pielke *et al.* (2007) argue that global population at risk from malaria would increase by 100 per cent by 2080 without taking climate change into account, whereas accounting for climate change would further increase this risk by at most 7 per cent. Hence, it is important to design adaptation policies that can be described as win-win strategies—that is, strategies that provide benefits for multiple threats and not focus exclusively on climate change.

The evidence presented in this paper suggests that: (a) climate change impacts are increasing over time indicating the increasing climate sensitivity of Indian agriculture; (b) the climate response function is robust to the inclusion of weather variability and annual weather deviation in the model specification and (c) accounting for spatial autocorrelation is important due to the presence of significant spatial clustering of the data. Further, the climate change impacts are significantly lower after incorporating spatial effects in the model specification. Among other things, inter-farmer communication could be responsible for the presence of spatial effects. Arguing that it can be viewed as effective adaptation strategy, the paper also investigated for the presence of interaction between farmers through focus group meetings at several villages in Andhra Pradesh and Tamil Nadu. The study findings generally support the notion of information diffusion between farmers, with diverse agents acting as the information providers. In contrast to the general belief that agricultural extension centres operate as primary source of information, the evidence from the field discussions indicate that in reality farmers are benefitted very little through these government outfits. The main sources of information are in fact the large farmers in the neighbourhood, fertiliser and pesticide dealers, seed providers and family members.

Given the importance of adaptation as suggested by these findings one issue needs careful attention: How to adapt and adapt to what?

How to Adapt and Adapt to What?

The impact assessment literature mainly focussed on what could be termed as engineering/technological adaptation options. One measure of the potential and cost of adaptation is to consider the historical record of past speeds of adoption of new technologies. For example, Reilly and Schimmelpfennig (1999) show the relative speed of adoption of various adaptation measures. While the time taken for relatively soft adaptation measures such as variety adoption and fertiliser adoption could be in the range of 3 to 10 years, the hard options like development of irrigation equipment and irrigation systems take much longer time. Jodha (1989) also provides similar estimates based on evidence from post-independent India. These adjustment times indicate that for effective implementation of adaptation strategies, appropriate planning must start well before the manifestation of climate change. Also, soft options could be more cost-effective and hence should be explored first. Often the soft options (which include enhancing inter-farmer communication discussed here) may provide dual advantage of gearing up for the future climate change as well as providing benefits under the present-day conditions.

This leads the discussion to the next issue: adapt to what? This has significant policy relevance in the ongoing discussion on 'mainstreaming' the climate policies. For vast majority of developing countries (including India) climate change is a distant and invisible threat whereas they are presently exposed to a range of stresses (including climate related shocks such as cyclones, droughts and floods). If climate change response strategies were to be embraced by these countries it is imperative that such response strategies are aligned with development agenda. Also, for adaptation to be employed and to be effective they would need to be shown to be relevant to local people and they would need to be integrated into decision-making structures and processes. It is unrealistic to expect special policy initiatives to deal with climate change adaptation by itself, especially when so many of the suggested adaptation measures (such as drought planning, coastal zone management, early warning etc.) are currently being addressed in other policies and programmes. Thus, as argued by Smit and Benhin (2004), in order to mainstream, it is essential that the analyst pay serious attention to: (a) the climate related issues that matter now to the community; (b) the management or coping strategies presently employed by the local community to deal with those conditions and (c) the policy structure that exist now to deal with such conditions.

Underlying this is the implicit assumption that adaptation strategies geared to cope with large climate anomalies that society faces, currently embrace a large proportion of the envelope of adjustments expected under long-term climate change. In other words, the climate policies (at least in the local context) need not be something different from the development policies. However, this need not be interpreted as nullification of need for research on climate change-specific adaptation options. On the contrary, the two should be seen as complimentary to each other. The recommendations of Interagency Report (2003) are also on similar lines.

Studies of adaptation to current climate also make it clear that human activities are not always as well adapted as one would want them to be. As argued by Burton *et al.* (1993), the losses suffered due to climate extremes cannot be ascribed to the events alone because lack of appropriate human adaptation and sometimes maladaptation account for significant losses. In this context it may be worth noting the experiences with the Super Cyclone in 1999 that devastated the state of Orissa. There is a general agreement that cyclone devastation was worsened significantly by deforestation on the coast. Satellite pictures show that 2.5 square kilometres of mangroves were lost in the 1970s, every year. In a recent study, Das (2007) has rigorously established that about 92 per cent of the deaths could have been avoided at Kendrapada during 1999 Super Cyclone in Orissa had the mangrove forests that existed in the year 1950 were still in place. Similar savings would have also been possible for animal lives and material losses. Enhancing the inter-farmer communication—suggested as adaptation strategy in this paper—could also lead to improvement in the present coping capacity of farmers.

References

Burton, I., R.W. Kates and G.F. White (1993). *The Environment as Hazard.* Second Edition. New York: Guilford Press.

Cline, W. (2007). *Global Warming and Agriculture: Impact Estimates by Country*. Washington D.C.: Peterson Institute.

Das, Saudamini (2007). "Valuation of the Storm Protection Services of Mangrooves of Orissa", *SANDEE Working Paper* No. 25-07.

Fussel, H.M. and R.J.T. Klein (2006). "Climate Change Vulnerability Assessments: An Evolution of Conceptual Thinking", *Climatic Change* 75(3): 301-329.

Interagency Report (2003). *Poverty and Climate Change: Reducing the Vulnerability of the Poor through Adaptation*. Interagency report by the African Development Bank, Asian Development Bank, Department for International Development (UK), Federal Ministry for Economic Cooperation and Development (Germany), Ministry of Foreign Affairs—Development Cooperation (the Netherlands), Organisation for Economic Cooperation and Development, United Nations Development Programme, United Nations Environment Programme and the World Bank.

Jodha, N.S. (1989). "Potential Strategies for Adapting to Greenhouse Warming: Perspectives from the Developing World", in N.J. Rosenberg, W.E. Easterling, P.R. Crosson and J. Darmstadter (eds.), *Greenhouse Warming: Abatement and Adaptation, Resources for the Future*. Washington, DC. pp.147-158.

Kolstad, C. (2000). *Environmental Economics*. New Delhi: Oxford University Press.

Kumar, K.S. Kavi (2003). *Vulnerability of Agriculture and Coastal Resources in India to Climate Change*, Research Report submitted to Ministry of Environment and Forests. New Delhi: GoI.

Kumar, K.S. Kavi and J. Parikh (2001a). "Socio-economic Impacts of Climate Change on Indian Agriculture", *International Review of Environmental Strategies* 2(2): 277-293.

———. (2001b). "Indian Agriculture and Climate Sensitivity", *Global Environmental Change* 11(2): 147-154.

Mall, R.K., R. Singh, A. Gupta, G. Srinivasan and L.S. Rathore (2006). "Impact of Climate Change on Indian Agriculture: A Review", *Climatic Change* 78: 445-478.

Mendelsohn, R., A. Dinar and A. Sanghi (2001). "The Effect of Development on the Climate Sensitivity of Agriculture", *Environment and Development Economics* 6(1): 85-101.

Mendelsohn, R., W. Nordhaus and D. Shaw. (1994). "The Impact of Global Warming on Agriculture: A Ricardian Analysis", *American Economic Review* 84(4): 753-71.

Pielke, R. Jr., G. Prins, S. Rayner and D. Sarewitz. (2007). "Lifting the Taboo on Adaptation", *Nature* 445: 597-598.

Polsky, C. (2004). "Putting Space and Time in Ricardian Climate Change Impact Studies: Agriculture in the US Great Plains, 1969-1992", *Annals of the Association of American Geographers* 94(3): 549-564.

Reilly, J.M. and D. Schimmelpfennig (1999). "Agricultural Impact Assessment, Vulnerability, and the Scope for Adaptation", *Climatic Change* 43: 745-788.

Schlenker, W., W.M. Hanemann and A.C. Fisher (2005). "Will US Agriculture Really Benefit from Global Warming? Accounting for Irrigation in the Hedonic Approach", *American Economic Review* 95(1): 395-406.

Smit, B. and J. Benhin (2004). "Tools and Methodologies for Mainstreaming Vulnerability and Adaptation to Climate Change into Sustainable Development Planning", *www.unep.org/themes/climatechange/PDF/Paper_No.5.pdf*

Solomon, S., D. Qin, M. Manning, Z. Chen, M. Marquis, K.B. Averyt, M. Tignor and H.L. Miller (eds.) (2007). *Climate Change 2007: The Physical Science Basis*. Contribution of Working Group I to the Fourth Assessment Report of the Intergovernmental Panel on Climate Change. Cambridge, United Kingdom and New York: Cambridge University Press. pp.996.

Stern, N. (2006). *The Economics of Climate Change: The Stern Review*. Cambridge: Cambridge University Press.

12

Biotechnology in Indian Agriculture

Evidence from Panel Studies on Bt Cotton

N. Chandrasekhara Rao

Technical change in agriculture increases production at the same level of input use and postpones the operation of the law of diminishing returns. It increases production at reduced unit costs/prices in real terms. The introduction of seed-fertiliser technology in early 1960s increased total factor productivity in Indian agriculture significantly. This, along with infrastructure development and interventions in price setting and marketing, has played an important role in achieving self-sufficiency in food grain production. However, there is empirical evidence to believe that growth of total factor productivity has come down since late eighties.[1] In fact, by the end of eighties, the area under modern varieties has already reached the ceiling levels particularly in the frontline states (Kumar and Mruthyunjaya, 1992; Kumar *et al.*, 2008) and major crops and yield levels reached a plateau. At the all-India level, the yields of most of the crops have either stagnanted or declined. Very few crops registered an increase (Table 12.1). The availability of tools of modern biotechnology especially transgenics assume importance in this situation.

These tools provide an opportunity to infuse a new round of technology into Indian agriculture. However, the use of these technologies is surrounded by many controversies, many of them ill-informed (Stone, 2002). Public reservations and uncertainties about impacts add to the danger that appropriate biotechnologies will remain inaccessible to those who stand to benefit the most (Lipton, 2001). Therefore, there is a need in this situation for a clear understanding of the nature of these technologies,

1. Please see Kalirajan and Shand (1997) and Kumar and Rosegrant (1994).

their uses, pattern of product development and commercialisation, impact of the commercialised products on yield, profits and on different groups of farmers. This paper attempts to address these issues in the context of Indian agriculture.

Thus, the paper is organised as follows. Section II brings out an overview of agricultural biotechnology with concept, role and benefits. Section III explains the pattern of commercialisation and Section IV gives the social impact and issues of concern arising with the availability of biotechnology. Section V presents the findings of two longitudinal field surveys for Bt cotton conducted by the authors to examine the profitability across different sections of farmers and the last Section concludes with policy suggestions to utilise these technologies for achieving the goal of poverty reduction.

Overview of Agricultural Biotechnology

The term biotechnology covers a wide range of scientific techniques and products that can be used in numerous ways to boost and sustain the productivity of crops, livestock, fisheries and forests. It is the technique of using living organisms or their parts to make or modify products, improve plants or animals or develop micro-organisms for specific use. Modern agricultural biotechnology includes a range of tools that scientists employ to understand and manipulate the genetic make-up of organisms for use in the production or processing of crops, livestock, fisheries and forestry.[2] The modern biotechnology broadly has two groups of technologies. The cellular approaches include tissue culture, which is used in micro propagation and animal reproduction. The molecular approaches include genomics and bioinformatics, diagnostic procedures, molecular markers technology[3] and genetic engineering. Other biotechnologies include use of biofertilisers and biopesticides. Genetic modification (GM) (sometimes called genetic engineering) refers to approaches within the broad domain of biotechnology, which are distinguished from other biotechnology

2. FAO (2004) and Herdt *et al.* (2007) explain in detail the subject matter and different techniques in agricultural biotechnology.
3. Reece and Haribabu (2007) bring out the utility of marker-assisted selection (MAS) in poverty reduction by developing improved crop cultivars.

techniques by allowing the transfer of genes between different organisms.[4] It is important to keep in mind that genetic modification is a subset of biotechnology and also that it is a method, which works with the support of non-genetic modification biotechnology methods as well as conventional methods.

Table 12.1

Levels of Yield of Major Crops in India since 1950

(in kgs per hectare)

	1950-1954	1955-1959	1960-1964	1965-1969	1970-1974	1975-1979	1980-1984	1985-1989	1990-1994	1995-1999	2000-2005
Rice	772	866	1016	980	1105	1207	1352	1584	1807	1898	1984
Wheat	724	728	834	1040	1294	1456	1770	2066	2387	2608	2669
Sorghum	428	464	508	506	513	681	698	708	826	834	777
Pearl millet	556	594	598	606	554	466	523	476	636	712	850
Maize	712	806	976	1012	1036	1074	1255	1297	1525	1723	1882
Finger millet	na	na	na	na	867*	1042	1036	1084	1255	1374	1495
Gram	na	na	na	na	570*	658	657	679	755	791	785
S. cane	32502	34974	44418	45818	49758	51784	56630	61388	66737	69418	64405
G. nut	724	760	752	676	763	838	853	896	948	1033	1055
Sesamum	na	na	na	na	176	197	232	259	302	318	351
Cotton	94	98	118	114	138	153	164	198	241	234	258
Soybean	na	na	na	na	761*	889	721	725	937	1071	957
Sunflower	na	na	na	na	671*	563	522	407	556	557	530
RS & mustard	na	na	na	na	1353*	1450	1491	1626	1771	1820	1646
Tea	na	na	na	na	612*	495	626	772	874	886	1027

Note: * Relate to only 1974-75.

Source: DES, Government of India.

The biotech gene revolution in agriculture is said to follow from the successes of the green revolution. Unlike the green revolution model, where a simple science breakthrough (high yielding dwarf varieties) could be transferred in an essentially directed and top-down way through a vast state infrastructure, in biotech the innovations are more complex and the routes to application less obvious. In fact, biotechnology is a collection of diverse

4. The organisms resulting from genetic modification are termed variously as genetically modified organisms (GMOs), and living modified organisms (LMOs). The crops so modified are called genetically modified crops and also transgenics as the genes are moved across crops.

and reinforcing enabling technologies with a wide range of applications in agriculture, forestry, food processing, waste management, pollution control, chemicals, raw materials, energy, cosmetics, pharmaceuticals and probably other sector that will become apparent in the future. Some more difference in these two technologies are brought out in Table 12.2.

There is unanimity of opinion on the potential of biotechnology to assist in the development of agriculture (FAO, 2004; GM Science Review Panel, 2003; Rao, 1994; Rao and Gulati, 1994; Middendorf *et al.*, 1998; Damodaran, 1999; Kalaitzandonakes, 2000; Qaim, 2001; Rangarajan, 2002; Scoones, 2002a and 2002b; UNDP, 2001; World Bank, 2007). It can break the yield barriers not possible to achieve with the conventional breeding methods, close the yield break by developing products to overcome biotic and abiotic stresses, and also sustain the present yield levels by enabling technologies that can withstand climate change and degradation of natural resources including water and soil. Apart from these, these technologies can also play an important role by tailoring technologies to increase the value addition like higher amino acids in soybean, Vitamin A rich rice often called golden rice, protein rich potato etc., to increase the quality of the produce. A unique feature of this technology is that it can reduce the usage of chemical inputs like pesticides and therefore can be beneficial to the environment. However, it depends on the way its potential is harnessed.

The green revolution technologies have not made the same level of impact on dryland agriculture as it did in the case of well-endowed regions. However, dryland agriculture continues to play a very important role in the total food and agricultural production. Dryland agriculture in India covers 67 per cent of the net cultivated area and currently accounts for more than 60 per cent of food grains (90-95 per cent of millets), almost 80 per cent of oilseeds, 90 per cent of green legumes and 70 per cent of cotton. Even 50 per cent of paddy is grown under rainfed conditions. Increasing the productivity of dryland agriculture becomes absolutely essential also for the reason that only the improvement in the productivity of dryland agriculture can lead to benefits of growth being shared by many (Rangarajan, 2002). The tools of biotechnology, by their nature, are useful in developing technologies to suit these marginal environments. However, much depends on how these are harnessed.

Table 12.2

Biotechnologies versus Green Revolution Technologies

Item	*Green Revolution Technologies*	*Biotechnologies*
Drivers	It is the discoveries in the field of chemistry that paved the way for development of seed-fertiliser technologies.	The developments in the field of biology have driven the biotechnologies.
Domain	Public sector.	Private sector with proprietary rights.
Nature of technology	Slow, not precise. Often the resultant variety inherits some undesirable characteristics.	Quick and precise. Can target the required characteristic by identifying the gene.
Dependencies	Called seed-fertiliser technologies, as the seeds are developed to respond to fertilisers. Availability of irrigation is crucial to use of these technologies. Therefore, dependence on chemical inputs increases.	Seed is central to this technology. So, no dependence on chemical inputs.
Access	Mainly developed to reach well-endowed regions. But later, extended to drylands also. But not with great success in marginal environments.	They can be used to tailor technologies for marginal environments also. These also can be an answer to the challenges posed by climate change.
Coverage	Encompasses crops only.	Beside crops, improvements in livestock, fisheries and forestry are possible.
Nature of crops	Started with staple crops like wheat and paddy. Later slowly spread to other crops.	Started with cash crops like soybean, maize, cotton and canola. It is in principle possible to extend these technologies to other crops also.
Nature of technology	Scale neutral. But, not resource neutral. The technology development is continual.	Can be scale and resource neutral. But the method of harnessing the technology becomes crucial. The technology development is continuous because of the rapid pace of discovery and it can be obsolete in a quick time.

Pattern of Commercialisation

The commercialised biotech products in agriculture so far tried to address the issue of reducing the yield gap caused by biotic stresses like insect pest damage. The other tasks in crop improvement like breaking the yield barriers (Figure 12.1) are not targetted so far. The first generation biotechnology products in agriculture have been crops with improved agronomic properties, such as herbicide tolerance and resistance to particular insect pests. Technologically, these are rather elementary, scientific 'easy pickings'. They require over expression of a single gene coding for one enzyme or toxic protein. They provide various input traits, whereas the second generation bioengineered crops incorporate different desirable output traits. Second generation bioengineered crops with enhanced quality traits, such as corn with high oil and lysine content, hybrids with increased levels of amino acids, healthier oils in soybean, neutraceuticals—blending regular food product with health enhancing attributes like golden rice etc., are being developed, targetting the food, feed and edible oils markets. These are technically more complex than first generation traits. They are associated with many genes or gene complexes acting in concert. A direct implication is that to advance second generation agrobiotechnologies, systematically, researchers will have to learn how to coordinate expression of several genes over the growing period of a crop. In this direction, plant genomics—the discovery and the study of many genes (simultaneous)—is expected to provide solution.

Figure 12.1

Three Main Tasks for Improving Crop Yields

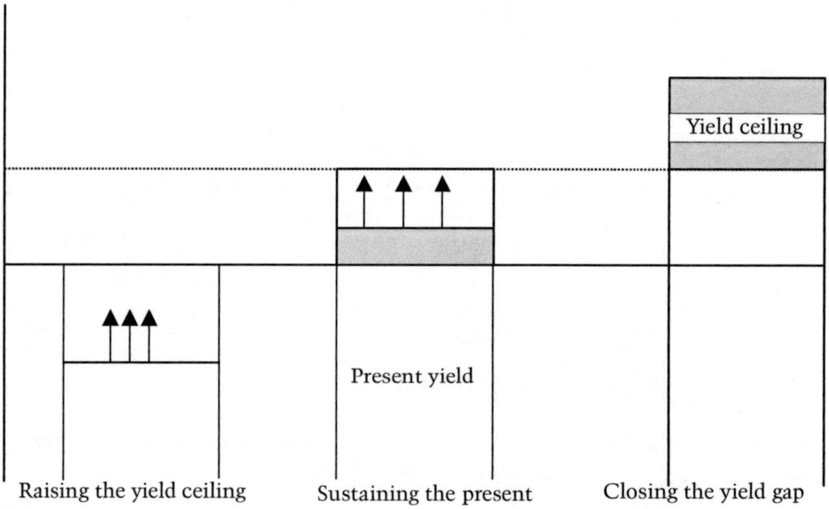

Source: Plucknett (1993).

The emergence of genomics and bioinformatics in the late 1990s radically changed the research paradigm in biotechnology and elevated expectations about the rate of discovery and product development in agro-biotechnology. These developments in genomics also alter the research method and scope in plant biotechnology. For the first two decades, the focus of plant biotechnology was on the isolation of commercially useful genes and on the methods of inserting and expressing such genes in commercially relevant species. Genes will be in excess over the next several years. The emerging research paradigm is the study of complex biochemical and physiological problems as an integrated system, rather than on a gene-by-gene basis (Kalaitzandonakes, 2000).

The stiff opposition to imports and permissions for cultivation of genetically modified crops in the world for the past few years seem to be easing since 2004. The European Union has allowed import of genetically modified maize and also permitted growing transgenics during the year, ending four years of moratorium after the United States of America approached the appellate body of the World Trade Organization. Therefore, the area under transgenics started increasing at a higher rate after that. On the other hand, the extension of transgenics area to food crops is long overdue.

Figure 12.2

Commercialisation of Transgenics in the World

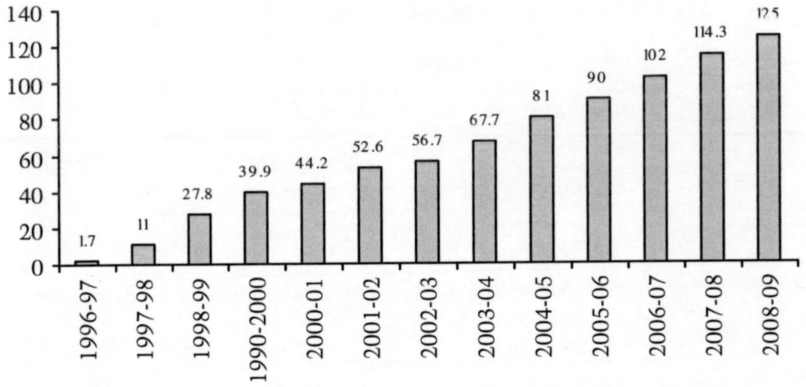

Source: James (2008).

The transgenic crops were commercialised in 1996 in the world. The area has been increasing steadily at a double-digit growth rate per annum and the area reached 125 million hectares in 2008 in 25 countries (Figure. 12.2). In 2008, USA (62.5 mha), Argentina (21.0 mha), Brazil (15.8 mha), India (7.6 mha), Canada (7.6 mha) and China (3.8 mha) account for 95 per cent of the total area under transgenics. When we look from the point of view of traits commercialised, nearly 80 per cent of the commercialised area is under herbicide tolerance (Figure 12.3). Cotton is the only crop in India that has been commercialised. The area under transgenic cotton in the country increased at a rapid rate to reach 7.6 million hectare in 2008-09 from a negligible 45,000 hectares in 2002-03 (Table 12.3).

Figure 12.3

Traits Covered under Transgenic Crops in 2008

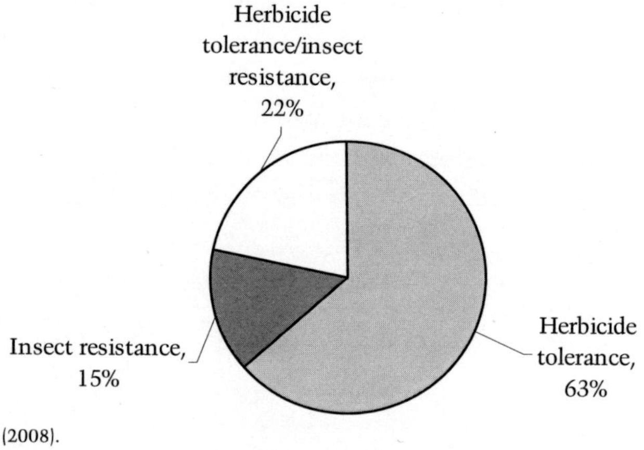

Source: James (2008).

Table 12.3

Commercialisation of Transgenic Cotton in India

Year	*Area under Bt Cotton (ha)*	*Per cent of Total Cotton Area*	*No. of Farmers*
2002-03	44,500	0.58	54,000
2003-04	1,00,000	1.31	-
2004-05	5,00,000	5.57	3,00,000
2005-06	13,00,000	14.38	10,00,000
2006-07	38,00,000	41.27	23,00,000
2007-08	62,00,000	68.88	38,00,000
2008-09	76,00,000	82.00	50,00,000

Source: James (2008).

In India there are more than 200 biotech cotton hybrids from different private companies at present. The resistance to bollworms is the trait in all these hybrids. Majority of these companies source this transgene from multinational companies (MNCs), though few also developed indigeneously. Now, Bollgard II has come incorporating more genes into cotton hybrids enabling resistance to all bollworms. Though cotton continues to be the only crop with released transgenics, field trials are on in several crops. Some biotech hybrids in brinjal may come out in near future and herbicide tolerant maize may be released in a few years by private companies. Field trails are going on and in different stages for 13 crops, at the end of 2005—in brinjal, cotton, cabbage, corn, groundnut, pigeon pea, mustard, potato, sorghum, tomato, rice, okra, cabbage and cauliflower. However, on the whole, most of the research in biotechnology in the country is in the private sector and is concentrate heavily on Bt hybrids in cotton.

Social Impact and Issues of Concern

The impact of any agricultural technology on the society can be direct and indirect.[5] The direct effect is on the holding farmers through increased productivity and reduced costs thereby improving their welfare. The indirect effect is through reducing the prices of food; creation of employment and rise in wages; thereby positively impacting employment, wage and income effects in other sectors of economic activity. The first and most desirable way is by reducing the cost of food by increasing productivity. This can benefit net purchasers of food in rural areas and also urban poor. This will also benefit farmers by increasing the net profit with decreasing unit cost of production. The green revolution depended on this strategy for agricultural development, which is subservient to the overall objective of poverty reduction. This is in contrast to the reform strategy in agriculture which aims at bringing dynamism to agriculture sector by making relative prices *viz.*, terms of trade favourable to agriculture, wherein the net purchasers of food will suffer.

The new technologies can result in reduction of poverty if they increase the opportunities for employment, as a section of the farmers *viz.*, small and marginal farmers also earn through wage employment, besides

5. An excellent analysis can be seen in de Janvry and Sadoulet (2002).

the agricultural labourers. The higher rate of growth in agriculture will also spur non-farm activities in the rural areas leading to higher employment avenues in rural areas and this higher growth rate in agriculture can also relax the wage good constraint on industrial growth. Another direct way is benefitting the small and marginal farmers by making improvement in the crops they grow and increased uses for these crops.[6]

Both the direct and indirect effects are important in a country like India in view of the higher poverty rates among the farmers,[7] higher rural poverty and concentration of poor[8] in rural areas, and higher proportion of net purchasers in rural areas.[9] How far the biotechnologies can lead to either of these effects? Using the technology instrument as part of a strategy for poverty reduction requires careful *ex-ante* analysis of how the nature of technology, the nature of poverty and the economic and institutional context in the particular region where technology is released affect the distribution of benefits and losses (de Janvry and Sadoulet, 2002). Therefore, we try here to see the nature of technology in brief to understand the likely impact on poverty. The emerging literature on biotechnology shows that poverty reduction effects may not be very high with the way the research is organised now. This is despite the fact there is unanimity on the fact that this technology has the potential for poverty reduction and these tools are much more powerful than the green revolution technologies. We try to see the problems with the biotechnology product development in brief here.

The biotechnology research is dominated by few life science multinational corporations and in the domain of private sector. Their interest is in maximising profits on their research by concentrating on

6. Like for example, biofuels and applications in food processing. Fan and Hazell (2000) also argue that the policymaking in India should concentrate on marginal areas and improve these crops.

7. The poverty among farmers is 30.73 per cent compared to a rural average of 23.99 and 28 per cent higher according to the National Sample Survey Organisation data analysed in Bhalla (2006).

8. The rural poverty in 2004-05 was 28.3 and higher compared to urban poverty, which was 25.7 (Dev, 2008). His analysis also brings out that 72 per cent of total poor and 67 per cent of extreme poor live in rural areas.

9. Dev and Ranade (1998) showed how the net purchasers of food are higher in proportion in the rural areas of the country.

technologies in crops, which can offer largest possible and lucrative market. Therefore, if left to markets, these technologies may benefit high value crops and well-endowed regions. This is amply evident from the present state of research and commercialisation of transgenics discussed above. There are no initiatives as far as rice, wheat, oilseeds, pulses and minor millets are concerned. The country's agriculture is dominated by rainfed agriculture. Even if the potential of irrigation is achieved by completing all the feasible projects, nearly half of the cultivable area would still be under the mercy of rain gods. The majority of the small and marginal farmers cultivate the so-called orphan crops. Unless efforts are made to develop varieties that can increase the genetic potential of these crops in dryland conditions, the goal of poverty reduction cannot be addressed effectively.

The MNCs undertaking research in agricultural biotechnology have been acquiring seed companies around the globe as a part of horizontal and vertical integration leading to concentration in seed market. Faced with the decline of profitability in the agrochemical sector, these companies started complementary product development like herbicide-resistant hybrids. Because of this complementary product development, seed became a package now (Rao, 2004). This development may actually increase chemical use, intensify damage to environment and affect livelihoods of the poor in ecologically fragile areas, at a time when the harmful effects of overuse of chemicals in agriculture are recognised. More than that, weeding is a manual operation done mostly by women in agriculture and its replacement by chemicals would only reduce employment opportunities and may increase destitution in villages. The biological protection of the seed from being reused has been increasing over time with the seed producer firms trying to do away with open pollinated varieties since the past two decades. This process is likely to gather momentum and lead to increase in seed replacement rates and higher per cent of cost on seed in the total cost of cultivation than before.

These technologies are also said to have solutions to the problem of malnutrition, quite serious in developing countries like India. Some of the applications like golden rice and protein rich potato (Protato) are shown as evidences in this regard to alleviate deficiency of vitamin A and protein, respectively. Food fortification through transgenic route is difficult because the availability of vitamin A, protein etc., also depends on the availability of other elements. Thus, it should be viewed as complement to existing interventions (Dawe *et al.*, 2002). However, as the pace of discovery being very fast in the modern biotechnology, several new applications can be done to address the problem of malnutrition more effectively in times to come.

The use of low-end biotechnological tools like biofertilisers,[10] biopesticides,[11] tissue culture, molecular markers etc., should not be neglected in the excitement created with genetic modification. The problem of biotic and abiotic stresses in agriculture can be solved by conventional plant breeding techniques, low-end biotechnological tools and then genetic modification. There is a need for a rigorous cost-benefit analysis between these alternative techniques for each of the biotic and abiotic problems, before they can be commercialised. The limited investment for agricultural research needs to be allocated between conventional and biotechnological research based on the analysis of short and long-run returns. On the whole, the proportion of our agricultural research expenditure on biotechnological research is very small compared to developed countries.[12]

The commercial applications of biotechnology need to keep in mind the preservation of biological diversity of the country and trading interests of the country. India is the primary and secondary sources of biological diversity in respect of number of crops like rice and the country should be kept transgenic-free to preserve the biodiversity, as many countries are doing. The exports for some of the crops like *basmati* rice and soyabean may be affected with genetic modification. Many European countries import soyabean as a non-genetically modified source. In future, genetic modification can also act as a potential sanitary and phytosanitary measure (Johnson, 2002). The importers can ask exporters to identify the source of origin from the producer to the consumer, which may be very difficult in the case of developing countries like India. The developing countries from

10. The efficacy of biofertilisers in improving the yield and reducing the need to use chemical fertilisers was proved beyond doubt in many studies (See for e.g. Ghosh, 2004).

11. Birthal *et al.* (2000) clearly brings out the effectiveness of biopesticides as a part of integrated pest management practices.

12. The proportion of agricultural research budget spent on biotechnological research is only 5-10 per cent in developing countries, compared to 16 per cent in developed countries (Pingali and Raney, 2005).

tropics have an advantage in case of certain crops because of the climate and they export these commodities to the temperate developed countries. The genetic modification offers opportunities for the temperate countries to evolve resistance to cold in these crops to be able to cultivate in temperate climate and substitutes for imports. This is likely to affect the exports, growth and employment opportunities in the developing countries.

The process for permitting commercialisation of these technologies needs to be transparent with the participation of end users in technology development to help in priority setting of demand based technology development.[13] The crux of this issue is that the tools of biotechnology represent a precise continuation of the plant breeding methods and also dramatic deviation from them. It has to be stressed that they are only tools and the ends have to be properly planned keeping the agroclimatic and socioeconomic needs of the host society. They are complementary to the conventional plant breeding (CPB) methods and not substitutes to CPB, integrated pest and nutrient management, livestock breeding, feeding and management systems. In fact, they can be harnessed better where the conventional research efforts are strong. To conclude, the biotechnology should be part of an integrated and comprehensive agricultural research and development programmes that gives priority to the problems of poor.

The entry of public sector in a big way is likely to alter the scenario in favour of the small farmers. They can release varieties in transgenics also as is being tried in cotton, instead of hybrids to provide scope for the farmers to reuse the seed for some years. This was really done in China in case of cotton and few other crops.

Performance of Bt Cotton

It would be worthwhile to examine in some detail the performance of Bt cotton as it is the only crop being commercialised so far in the country and also being highly contested. Therefore, an *ex-post,* farm level study on the impact of Bt cotton introduced in the country is highly relevant. The data from two surveys conducted by the authors in different agroclimatic zones of Andhra Pradesh on the performance of Bt cotton in two different

13. Some initiatives like Andhra Pradesh Netherlands Biotechnology Programme (APNLBP) in Andhra Pradesh in fact provide some good examples for participation of stakeholders in technology development. See Krishna and Reddy (2005) for details.

agricultural years are presented here. After a brief review of literature on Bt cotton, performance of the methodology of the field surveys, results from the two surveys, production function analysis for yields, benefits across social and size categories of farmers and impact of the technology on employment are presented in this section.[14]

A Literature Survey on Bt Cotton Impacts

The field trials of Bt cotton in India almost a decade back raised curtain for acrimonious debates on the utility of biotechnology in general and Bt cotton in particular. A survey of peer-reviewed articles in reputed journals clearly shows that this biotech cotton could effectively resist bollworm, reduce the spending on plant protection, and increase yield. The employment effect is negative in countries where it is harvested with machines as it reduces human labour utilisation for spraying of chemicals. But, the net effect is ambiguous where cotton is harvested manually. The quantum of these impacts varies across countries. Now, a brief review is attempted hereunder.

Qaim (2003) reported yield increase of 80 per cent using trial plot data. There were also refutations of claims of cost reduction and yield increase by some scholars. Based on field trials conducted in India, Shiva *et al.* (1999) concluded that the yield in all the trial plots were found to be low and a comparison of the local hybrid/variety cultivated and Bt showed that the yield from both the crops was more or less same. The cost of cultivation has also worked out to be same for all the farmers. They further concluded that wherever the bollworm pressure is not high, Bt cotton might not be economically suitable.

The first year of Bt cotton cultivation was a disaster in many of the cotton growing states like Andhra Pradesh, Maharashtra, Madhya Pradesh, Karnataka and Gujarat (Krishnakumar, 2003). In 2002-03, small saving in pesticide sprays (Rs 217/ac), lower profits and susceptibility to pink bollworm were reported in a study conducted in Maharashtra and Andhra Pradesh (Sahai and Rehman, 2003). Similar results with more attack of sucking pests were also reported from a season long study on Bt cotton in Andhra Pradesh (Qayum and Sakkhari, 2003). In another field survey in

14. The results presented are based on Rao and Dev (2009).

the first year of commercialisation in Tamil Nadu, Karnataka, Maharashtra and Andhra Pradesh, it was found that the yields increased by 34 per cent in Bt cotton over conventional cotton and that the farmers of Andhra Pradesh suffered a loss in average incomes (Naik *et al.*, 2005).

The performance in 2003-04 seemed to be slightly better compared to the previous year. In a study conducted in two districts of Maharashtra, it was found that the yield increased by 52 per cent in Bt cotton from 15.77 quintals per hectare in non-Bt cotton to 24.00 quintals in Bt cotton and the same study concluded that the profit per hectare was Rs 31,883 in Bt cotton as against Rs 17,797 in non-Bt cotton implying an increase of 79 per cent, though the cost of pest control increased slightly (Narayanamoorthy and Kalamkar, 2006). A nationwide survey by Nielson/ORG MARG for 2003-04 season concluded that there was a 60 per cent reduction in pesticide use and 29 per cent increase in yield leading to 78 per cent increase in net profit. The critics of Bt cotton questioned the objectivity of this survey, as Monsanto commissioned the survey. Further, these two studies did not take fixed costs into account while working out changes in costs and returns from Bt cotton cultivation over non-Bt hybrids. Qayum and Sakkhari (2004) made another season-long survey and negated these claims. The attack of bollworms in 2003-04 was also below normal. Therefore, they maintained that the performance in this year cannot be generalised. However, a study conducted in four of the main cotton growing states *viz.*, Andhra Pradesh, Gujarat, Maharashtra and Tamil Nadu concluded that the adopters of this new technology obtained 31 per cent higher yield and 88 per cent more profit than their counterparts growing conventional hybrids (Gandhi and Namboodiri, 2006).

It is also alleged that the cotton from the Bt hybrids is of inferior quality and that it cannot stand heavy rainfall conditions etc. (Sahai and Rehman, 2003). It is possible that the negative traits like short-stapled cotton, low yield etc., may be of the hybrids used to incorporate the Bt gene than the gene (Cry 1ac) itself. It is because the benefits of Bt technology in a given crop (B_i) is a function of genetic potential of the variety/hybrid (G_p) and the impact of transgene-Bt, which will be given by the cost reduction due to declined pesticide use and increased returns due to yield increase because of realising the yield potential by reducing pest damage as shown below.

$$B_i = f\,(G_p, Bt)$$

These technologies generally are expected to have a positive impact on employment, as there is always a huge underemployment and disguised unemployment in the country. The green revolution in the early 1960s introduced biological and mechanical technologies in agriculture. Several studies conducted after the green revolution found that the biological technology favoured employment creation and the mechanical technology displaced labour. As already mentioned earlier the Bt cotton cultivation is expected to increase labour use in India, though it is labour saving in countries like USA, where most field operations are mechanised. Therefore, it will be very important to analyse its impact on the use of human labour, with a particular emphasis on its gender implications.

Cotton is a very important commercial crop in the country with an area of around 9 million hectares. It accounts for 45 per cent of agro-chemicals used, though it is cultivated in only 5 per cent of gross cropped area. The loss due to insect pests is about 50 per cent of cotton output because some insects such as *Helicoverpa armigera* and whitefly have developed resistance to chemical pesticides (Birthal *et al.*, 2000). There were recorded suicide deaths of farmers in the last few years in Andhra Pradesh and Maharashtra due to failure of cotton crop because of pest menace. The productivity of cotton in the country is very low at five-year average of 204 kgs per hectare at the end of 2002-03. Therefore, any positive impact of this technology on pest control and yield stabilisation would be highly useful in improving the conditions of those involved in cotton cultivation. It can also give an indication of the potential of this technology for the entire agriculture sector, if significant positive impact on both these traits in cotton is proved beyond doubt.

Methodology

The present study follows double difference method combining the 'with and without' approach with 'before and after adoption' method. The study adopted multi-stage stratified random sampling method by taking one district from each of the four agroclimatic zones, where the Bt cotton hybrids were introduced. The sample selected in each agroclimatic zone was proportional to the area under Bt cotton in the respective zone. The selection of *mandals* was done based on the area under Bt cotton. We have chosen more than one *mandal* from each district to see that the sample is well spread out and more representative. The number of *mandals* selected for the study was nine from the four districts. The villages were selected based on the area under Bt cotton. The number of villages selected for the study was 14.

The farmers were selected after stratification based on farm size and social category. The farmers are stratified as small, medium and large based on size of owned landholding. The farmers owning less than 4.99 acres are considered small; who own landholding between 5 acres and 9.99 are considered medium and those who possessed 10.0 acres and more are considered as large farmers. The sample size is 623. The number of farmers from Warangal, Nalgonda, Guntur and Kurnool are 262, 188, 103 and 70 respectively. Primary survey was undertaken with pre-tested schedules and participatory methods like focus group discussions were used as supplementary. The data collected pertains to 2004-05 and to be specific *kharif* 2004-05. These same sample growers were surveyed again in 2006-07. As all the 186 non-Bt farmers in 2004-05 switched on to the new technology by the time of resurvey in 2007, 200 more farmers growing conventional hybrids are taken to represent the control group. The monetary values in both the years are deflated using the consumer price indices for agricultural labourers with 1986-87 for the state provided by the Indian Labour Bureau, while comparing these two.

Survey Results in 2004-05

The cost of production per acre is 17 per cent higher in Bt cotton at Rs 16,975 compared to Rs 14,507 for non-Bt cotton in the state and this difference is statistically significant (Table 12.4). This includes paid-out costs and imputed costs of depreciation, interest on owned fixed capital, rental value of owned land, family labour etc. The expenditure on insecticides decreased by 18.2 per cent in Bt cotton over non-Bt cotton. This decrease in cost of insecticides by Rs 594 is more than matched by increased costs on seed, labour costs, fertilisers and irrigation charges. All these changes are statistically significant except that in fertilisers. Out of the Rs 801 increase on labour, human labour accounted for the major portion *viz.*, Rs 676.

Table 12.4

Costs and Returns Per Acre in Bt Cotton and Non-Bt Cotton in Andhra Pradesh

(in Rs)

Item	*Bt*	*NBt*	*Per cent Change Over Non-Bt*
Casual labour	1780	1476	21**
Attached labour	218	127	71**
Family labour	1128	846	33**
Total human labour	**3125**	**2449**	**28****
Bullock labour	859	855	0
Machine labour	708	587	21**
Seed	1402	598	134**
Chemical fertilisers	1579	1603	-1
Manure	515	406	27**
Total fertilisers	**2094**	**2008**	**4**
Insecticides	2673	3267	-18**
Irrigation charges	98	54	83**
Interest on working capital	412	379	9
Miscellaneous	94	84	12*
Operational cost	**11466**	**10282**	**12***
Rental value of owned land	3608	2716	33**
Rent paid for leased-in land	753	523	44**
Depreciation	354	280	26**
Interest on fixed capital	794	706	12*
Fixed cost	**5509**	**4225**	**30****
Cost of production	**16975**	**14507**	**17***
Cost A1	10692	9716	10*
Cost A2	11445	10239	12*
Cost B1	12239	10945	12*
Cost B2	15847	13661	16*
Cost C1	13367	11791	13*
Cost C2	16975	14507	17*
Physical yield in quintals	9.49	7.21	32**
Cost A1/quintal	1127	1348	-16*
Cost A2/quintal	1206	1420	-15*
Cost B1/quintal	1290	1518	-15*
Cost B2/quintal	1670	1895	-12*
Cost C1/quintal	1409	1635	-14*
Cost C2/quintal	1789	2012	-11*
Average price per quintal	1750	1711	2
Farm business measures			
Gross income in Rs	16612	12338	35**
Net income	-363	-2169	83**
Farm business income	5166	2099	146**
Family labour income	765	-1323	158**
Farm investment income	4038	1253	222**

Notes: * and ** indicate significance at 1 per cent and 5 per cent, respectively.

Source: Field Surveys.

The reduction in insecticides is only 18 per cent, whereas Qaim and Matuschke (2004) showed from the review of studies that this reduction was to an extent of 77 per cent in Mexico and 65 per cent in China. In the study area, farmers sprayed pesticides with a fear of attack of Heliothis larvae from the adjoining fields as they are still not fully aware of the nature of Bt technology. This may change later and more savings in insecticides may result. Ismael *et al.* (2002) also observed that during the early stages of adoption, Bt growers use more insecticides than needed. Several studies (see for e.g., Qaim and Matuschke, 2004; Huang *et al.*, 2002) showed that the Bt technology adopting farmers also increase their input use. This also increases the cost of cultivation compared to non-Bt farmers. The study by Narayanamoorthy and Kalamkar (2006) also found a 34 per cent increase in the cost of cultivation of Bt cotton over non-Bt.

The physical yield obtained in Bt cotton is 9.49 quintals of seed cotton per acre compared to 7.21 quintals per acre for non-Bt cotton and 32 per cent higher than non-Bt and the difference is statistically significant. The coefficient of variation of yield reduced from 0.39 in the non-Bt cotton to 0.34 in Bt cotton showing that the variations across farms in terms of yield have come down in the farms growing Bt hybrids. This clearly shows the superiority of Bt cotton in increasing yields over non-Bt cotton and gives us a clue as to why the non-Bt cotton seed manufacturers are going out of business in the state. This result is particularly important in view of the very small average area per Bt adopter. It is only 2.77 acres in the state and it is basically a small farmer agriculture compared to other countries like USA, Argentina etc. For instance, in a field study in Argentina, the average area per Bt adopter was found to be 118 hectares in 2001 by Qaim and de Janvry (2005). It is well known that the economic theory states that the average costs matter in decision-making and deciding the profitability rather than absolute costs. The immediate fall-out of the higher yield in Bt cotton is that all its per quintal costs are lower over non-Bt cotton though the absolute costs are higher. The per quintal Cost A2, Cost B2 and Cost C2 are lower by 15 per cent, 12 per cent and 11 per cent, respectively in Bt cotton over non-Bt cotton (Table 12.4).

The net income *viz.*, gross income over Cost C2 is negative in both Bt and non-Bt cotton. But Bt performed better by being able to cover all the imputed costs along with paid-up costs except for a minor portion and improved over non-Bt by 83 per cent. The net income, farm business income, family labour income and farm investment income improved by 83 per cent, 146 per cent, 158 per cent and 222 per cent, respectively over the non-Bt cotton. All these are statistically significant at five per cent level. This clearly shows that Bt cotton outperformed non-Bt cotton in regard to all the measures. The farmer entrepreneur must be covering all these costs if farming is to be termed profitable. If he/she could not cover all costs, then covering the paid-up costs (Cost A1/A2) is crucial to remain in business. The farm business income, which shows us the excess of gross income over variable costs (Cost A2), is Rs 5,166 per acre in case of Bt cotton and 146 per cent higher comparatively.

The Commission for Agricultural Costs and Prices (CACP) reported negative net returns of Rs 656 per acre in 1996-97 in cotton in Andhra Pradesh, when the realised price per quintal of seed cotton was Rs 1,707. The cotton farmers from Punjab are also reported to have got negative returns in 1995-96 and 1996-97 (GoI, 2000). According to the Commission, in 2001-02, the net returns are only Rs 120 per acre, when the realised price is Rs 1,848. In Gujarat, for 2001-02, the net returns are Rs 1,531 per acre. In the same year, cotton farmers from Maharashtra and Tamil Nadu also are reported to have got negative returns (GoI, 2005). Therefore, it is not uncommon for the cotton farmers to get negative net returns, if all the paid-out and imputed costs are properly accounted as is done systematically by the CACP.

There may be two reasons for the negative net income for both Bt and non-Bt farmers. The first one is the fact that the actual rainfall in 2004-2005 in the study area is 33 per cent lower than the normal. This might have reduced the physical yield. The other factor is the drastic decline in prices for raw cotton during the year. It is noteworthy that the farm harvest price of cotton declined drastically in this year by 21 per cent compared to the previous year and is 13 per cent lower than the previous seven-year average of farm harvest prices. All this might have reduced the profitability of cotton farmers including Bt cotton farmers.

In a study conducted in two districts of Maharashtra, it was found that the profit per hectare was Rs 31,883 in Bt cotton as against Rs 17,797 in non-Bt cotton implying an increase of 79 per cent, though the cost of pest control increased slightly (Narayanamoorthy and Kalamkar, 2006). A nationwide survey by Nielson/ORG MARG for 2003-04 season concluded that there was a 78 per cent increase in net profit. In this background, the results of the study seem quite justifiable in the Indian conditions.

Economic Impact after Adoption of Bt Cotton in 2006-07

The results from resurvey in 2006-07 presented in Table 12.5 confirm the results obtained in 2004-05 regarding the significant yield advantage with Bt cotton. The major findings are:

- There was a 42 per cent increase in yield after the adoption of Bt cotton for the non-adopters in 2004-05.
- The adoption reduced the use of chemical insecticides considerably *viz.*, to an extent of 56 per cent compared to an 18 per cent reduction in 2004-05. This was made possible because of the rising awareness of the farmers as they continue to cultivate Bt hybrids compared to the initial stages of adoption. The adopters in 2004-05 were found to be applying with anxiety of pest attack. Now, it seems they have understood that there is no need to spray for bollworms up to a certain time period. This compares very well with several other countries, as brought out by Ismael *et al.* (2002).
- The cost of cultivation after adoption remained more or less similar compared to 30 per cent increase in 2004-05. As a result the Cost C2 per quintal has declined by 31 per cent after adoption in 2006-07 compared to an 11 per cent reduction in 2004-05.
- The net income became positive after adoption in 2006-07 and improved by 2.5 times after covering all direct and indirect costs.
- The farmers gained a farm business income of Rs 9,596 per acre after adoption of Bt.

The pattern of yield distribution for the farmers before and after adoption of Bt cotton is given by the non-parametrically estimated density functions for yield of cotton for the sample farmers in Figure 12.4 following Qaim (2003). It clearly shows the rightward movement of density function after adoption of the Bt by the same farmers in 2006-07. In fact, more than 70 per cent of farmers got less than 500 kgs per acre of cotton *(kapas)* yield before adoption and this situation changes altogether after adoption. Nearly 80 per cent of them got more than 700 kgs of cotton *(kapas)* yield after adoption of Bt. This brings out the reason why the adoption rates have been increasing so rapidly in the country, despite adverse media coverage and campaigns against the use and profitability in Bt cotton cultivation.

Table 12.5

Costs and Returns Before and After Adoption of Bt Cotton

(Per acre in Rupees)

Item	*Before (Non-Bt Cotton)*	*After (Bt Cotton)*	*Per cent Change*
Total human labour	2449	3249	18
Bullock labour	855	906	-6
Machine labour	587	886	34
Seed	598	897	34
Total fertilisers	2009	2103	-7
Insecticides	3267	1599	-56
Other costs	601	489	-19
Total operation cost	10282	10129	-12
Total fixed cost	4225	5927	25
Total cost	14507	16056	-1
Physical yield	7.21	10.27	42
Cost A1 per quintal	1348	922	-39
Cost A2 per quintal	1420	986	-38
Cost B1 per quintal	1518	1023	-40
Cost B2 per quintal	1895	1421	-33
Cost C1 per quintal	1635	1165	-37
Cost C2 per quintal	2012	1563	-31
Average price per quintal	1711	1920	0
Gross income	12338	19722	42
Net income	-2169	3667	251
Farm business income	2099	9596	307
Family labour income	-1323	5122	445
Farm investment income	1253	8141	478

Note: 1. The per cent change is worked out using the monetary values in constant prices.

2. The data from 2004-05 are used for 'before adoption' and that from 2006-07 are used for 'after adoption'.

Source: Field Surveys.

Figure 12.4

Estimated Density Functions for Yield

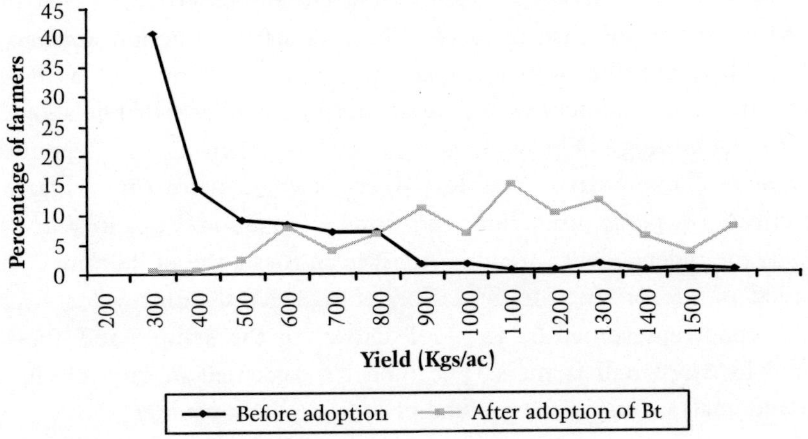

Source: Field Surveys.

As already discussed, any new technology of production is supposed to shift the cost function downward with the reduced costs for unit of output. The costs before and after the introduction of Bt are plotted against the Cost C2 in constant prices (Figure 12.5). It clearly shows that almost all the outputs are produced with lower level of cost with introduction of Bt hybrids in cotton compared to the conventional hybrids.

Figure 12.5

Cost Function Before and After Adoption of Bt Cotton

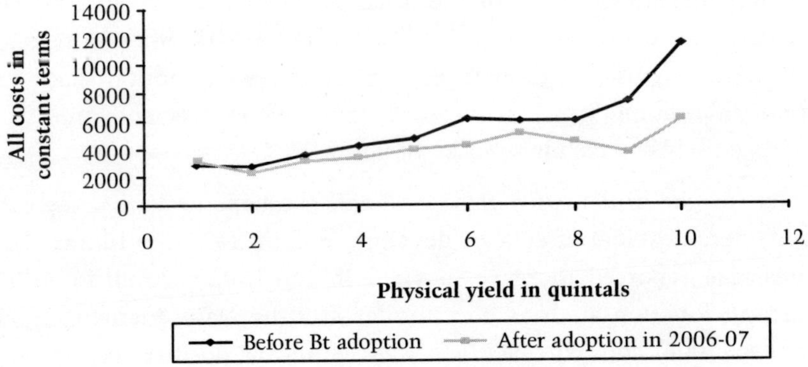

Source: Field Surveys.

Results of Production Function

Then, the question to be answered is whether the increase in yield is attributable to Bt technology alone or are there any other factors that may be leading to the increase in yield? The production function analysis is used to bring out the precise impact of Bt cotton technology as it can control for any adjustment in the input mix by the farmers while adopting the new technology. The yield in quintals is taken as the dependant variable and expenditure on fertilisers, expenditure on irrigation, expenditure on plant protection, education in years and age in years are taken as the independent variables. Apart from these, the Bt dummy is also included in the production function, along with a dummy for scanty rainfall zone represented by Kurnool district in the sample and another dummy for the small farmers. The model is specified in multiple linear form and analysed using the method of ordinary least squares (OLS).

The results show that there is significant impact of Bt cotton hybrid on the yield of the farmers (Table 12.6). The coefficient for Bt dummy turned out to be significant at 1 per cent level in both 'with and without' scenario in 2004-05 and 2006-07 as well as 'before and after' adoption scenarios. All the other variables contributed positively to the yield except dummy for scanty and erratic rainfall zone *viz*., Kurnool. The yields in this zone are significantly lower than the average yields and this difference came down from 30 per cent in 2004 to 8 per cent in 2006 mainly due to the better rainfall in that year. The small farmer dummy also turned out to be significant and negative implying that these farmers are getting 5 per cent lower yields than that for the other farmers in 2004. By 2006, the small farmers are also getting similar yields on par with other farmers and there was no significant difference. This endorses the notion that small farmers take some time to improve their awareness and ajust to a knowledge-intensive technology like Bt cotton.

The results of the study clearly validate the hypothesis that there can be considerable yield effects in developing countries like India. Other studies also reported increase in yield in the Indian conditions. The production function analysis of a similar kind by Narayanamoorthy and Kalamkar (2006) showed that there was significant positive impact on Bt cotton yield in their sample studies.

Table 12.6

Estimated Production Functions in Andhra Pradesh (n=623)

Item	With and Without Bt in 2004 (n=623)		Before and After Bt (n=367)		With and Without Bt in 2006 (n=814)	
	Coefficient	S.E	Coefficient	S.E	Coefficient	S.E
Constant	5.514*	0.433082	3.441*	0.634988	3.968*	0.382053
Bt dummy	2.826*	0.249625	3.915*	0.400266	4.234*	0.409716
Education	0.08297*	0.026825	0.0808**	0.033836	0.04337	0.024797
FYM (Rs/ac)	0.00008	0.000174	0.00216**	0.000944	0.00056*	0.00019
Fertilisers (Rs/ac)	0.00065*	0.000169	0.00397*	0.00082	0.00081*	0.000125
Pesticides (Rs/ac)	0.00023**	9.15E-05	0.00141*	0.000506	0.00031*	7.57E-05
Irrigation (Rs/ac)	0.00052**	0.000259	0.00408*	0.001566	0.00235**	0.001187
Kurnool dummy	-3.850*	0.369375	-1.320*	0.508671	-0.04295	0.401402
Small farmer dummy	-0.631*	0.228292	-0.485	0.314527	-0.330	0.25076
F		45.390		26.462		49.144
Adjusted R^2		0.363		0.358		0.321

Notes: * and ** indicates significance at 1 per cent and 5 per cent respectively.

Source: Field Surveys.

Following Qaim and Matuschke (2004), the positive yield effects can be explained in a damage control framework as below. If Y is the effective cotton yield, and F (.) is potential yield without insect damage, which depends on variable input, x., *D* (.) is the damage function determining the fraction of potential output being lost to insect pests. Crop losses depend on exogeneous pest pressure, *N*, and they can be reduced through application of chemical insecticides, *z*, and/or the use of Bt technology.

$$Y = F(x)\,[1 - D(z, Bt; N)]$$

If pest pressure is high and farmers use a lot of chemical insecticides in conventional cotton, Bt adoption should lead to substantial insecticide reductions. On the other hand, if there is uncontrolled pest attack with the chemical insecticides, Bt cotton could increase the effective yield by controlling the damage function, *D*. There is uncontrolled damage due to the bollworm attack in the study area, which is in fact true of entire country and this pest pressure is high as cotton is grown in a typically tropical climate in the state. Bt cotton cultivation in this situation resulted in increase in yield due to better management of the pest problem and thereby enabling to reach the potential of the variety/hybrid in question.

Benefits Across Social and Size Categories of Farmers

The biotechnology, as already stated at the beginning, is expected to benefit drylands and ecologically fragile areas. Then, what are the benefits of Bt cultivation to the farmers in these areas? The gains due to the new technology are ideally to be shared among all sections of the farmers equally. But, this does not happen in the real world due to differences in asset position as well as different positions due to social stratification in our society on caste lines. Therefore, one has to ask 'whether the new technology benefits the SCs & STs and small farmers on par with the OCs and large farmers?' At the least, are they getting any benefits from the technology to improve their financial position? We try to find answers to these questions by analysing the field survey data of the study here (Table 12.7 and Figure 12.6, 12.7 and 12.8).

Table 12.7

Per cent Changes across Different Categories of Bt Farmers over Non-Bt Cotton

Category	*Expenditure on Pesticides*			*Yield*			*Net Income*		
	With and Without in 2004	*After Adoption*	*With and Without in 2006*	*With and Without in 2004*	*After Adoption*	*With and Without in 2006*	*With and Without in 2004*	*After Adoption*	*With and Without in 2006*
Small farmers	-26.4	-56 *	-33*	10.09*	39*	49*	69	214*	221*
Medium farmers	-30.06 **	-65*	16	20.56*	33*	136*	90*	212*	730*
Large farmers	14.85 **	-38*	122	83.04*	93*	244*	120*	460*	211*
Irrigated farmers	-19.60 **	-62*	-23*	34.86*	34*	67*	149*	294*	306*
Rainfed farmers	-17.38**	-54*	5	28.08 *	32*	100*	46	185	180*
Warangal	-9.71	-59*	-31*	40.05*	41*	82*	139**	430*	249*
Nalgonda	-24.9 *	-54*	5	30.21*	16	101*	44**	92*	141*
Guntur	-29.5 *	-66 *	-9	18.85 *	49*	62*	19*	359*	884*
Kurnool	17.31	-7	4	85.85*	213*	62*	82	222*	200* times#
Total sample	-18.2*	-56*	-14**	31.62*	42*	80*	83*	251*	263*

Note: * and ** indicates significance at 1 per cent and 5 per cent respectively.

# indicates the net income increased from Rs 25 to Rs 4,480 per acre.

Source: Field Surveys.

It can be observed from Table 12.7 that all size categories of farmers, farmers from all the districts representing different agroclimatic zones, farmers from both irrigated and rainfed conditions could reduce their expenditure on pesticides, and increase the yield. The increase in yield over non-Bt is significant in most of the cases at one per cent level of significance. Therefore, it can be concluded that all categories of farmers got benefitted by cultivating Bt cotton. However, the net income increase is not significant in 2004-05 and also after adoption. This highlights the need for focussing biotechnological research on drought tolerance, so that farmers from drylands also get the benefits.

Then, the next question that needs an answer is that 'are there not any differences in the extent to which the benefits accrued to different groups of farmers?' Figures 12.6, 12.7, and 12.8 answer this question to some extent. We are not presenting the full data to save space, though these are available with us. It can be seen from Figure 12.6 that the farmers from Guntur, Warangal, irrigated category, large category and OCs got the maximum yield relative to the average in their respective groups. The farmers from the scanty rainfall zone *viz.*, Kurnool and Nalgonda obtained very poor yields and of course, the small farmers too including SCs. The similar situation can be observed in the case of net income also (Figure 12.7). The farmers from irrigated category, OC category, Guntur, Warangal and large category alone could cover all costs. All the others could not cover all costs, though they could get higher and positive farm business income over non-Bt cotton. However, as already explained, the deficient rainfall might have affected yield and net income in Nalgonda and Kurnool and also to other groups of farmers. These two districts basically fall in scanty rainfall zone, with the normal rainfall being 751 mm in Nalgonda and 670 mm in Kurnool compared to 993 mm in Warangal and 852 mm in Guntur. The area under irrigation in these two districts is also quite low with only 25 per cent of the Bt cotton area irrigated compared to a 54 per cent and 85 per cent irrigated areas in Warangal and Guntur respectively. However, the situation in terms of net income improved in 2006-07 with all the groups of farmers getting positive net income covering all the costs including those of family labour, rental value of owned land, depreciation, interest on owned fixed capital and working capital etc. (Figure 12.8).

Figure 12.6

Per cent Yield Levels of Different Bt Farmers Relative to Total Bt Sample

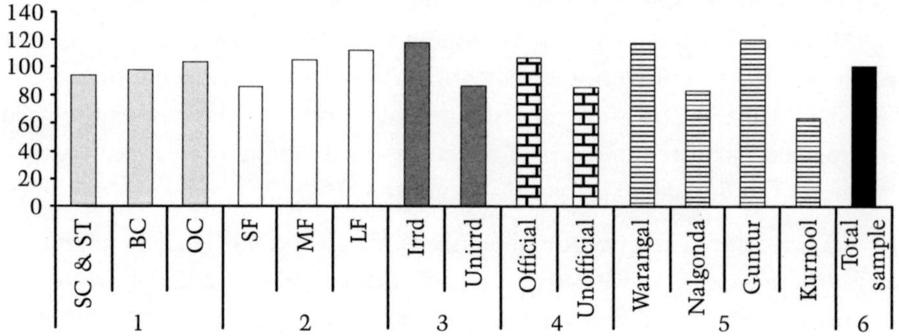

Source: Field Surveys.

Figure 12.7

Net Income from Bt Cotton to Different Categories of Farmers in 2004-05

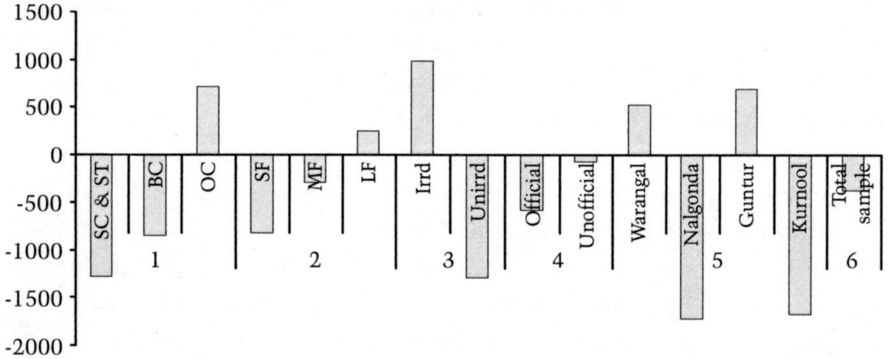

Source: Field Surveys.

Figure 12.8

Net Income from Bt Cotton in 2006-07

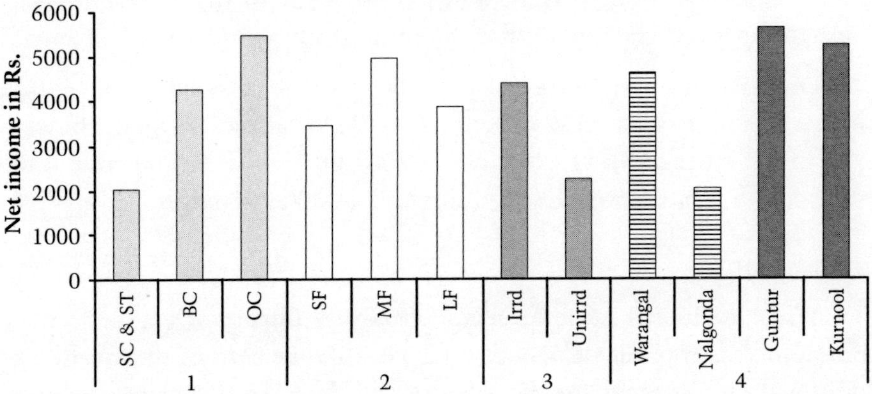

Source: Field Surveys.

Gains from the Technology and Sharing

The major concern regarding biotechnological applications by private companies is that the seed developer will appropriate all the benefits. Therefore, we tried to see the farmers' share of additional benefits across different countries. Across the developing countries, the farmers get major share and that is the reason for the faster adoption rates. Wherever their share is less as we can see in the case of Argentina, the diffusion is slow (World Bank, 2007).

Figure 12.9

Per cent Share of Benefit in 2004-05

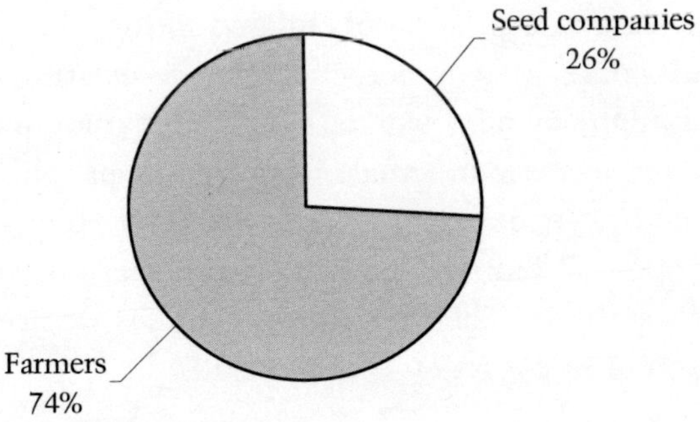

Source: Field Surveys.

Our field studies in Andhra Pradesh also showed that the farming community could get a major share of 74 per cent in 2004-05 (Figure 12.9). Later, as the cost of seed went down consequent to the intervention by state, the share reached more than 90 per cent in 2006-07, like in China.

Our estimations also reveal that there was a net gain to the cotton farmers to a tune of Rs 7,122 crore in 2006-07. We also estimate the benefits to be of the order of Rs 11,620 crore in 2007-08 and Rs 16,000 crore if the entire cotton area is covered with this cotton at 2006-07 prices.

Employment

While additional income generation is very important for the farming community as a result of new technologies like Bt cotton, the creation of additional employment for the labour is no less significant for the rural economy as a whole and landless labourers in particular. If the new technology increases the net income to the farmer and at the same time reduces the need for labour, the net effect on the rural economy may not be positive. Therefore, the impact of Bt cotton on employment is attempted here.

The total man-days equivalent (TMDE) for human labour (including casual, family and attached labour) for Bt cotton are 66.29 and are 21 per cent higher than non-Bt cotton and this difference is statistically significant (Table 12.8). Out of this increase, hired labour, family labour and attached labour accounted for 63, 27 and 10 per cent, respectively. While the major beneficiary of this increase is hired casual labour, there is also increase in the use of family labour and attached labour to a certain extent. The female labourers are the major beneficiaries among casual labourers. They got 10.05 days more of employment due to Bt cotton cultivation. On the other hand, the increase in utilisation of children in the labour force in Bt cotton is a cause of worry, as it is perpetuating the existing evil. The detailed operation-wise employment worked out in the study (not presented here) showed that the increase in female labour participation is due to increased utilisation mainly in harvesting and weeding. However, the results of resurvey in 2006-07 did not confirm this increase in employment in the after adoption scenario. Therefore, further studies are needed on the impact of introduction of Bt cotton on employment.

Table 12.8

Human Labour Utilisation per Acre in Bt Cotton and Non-Bt Cotton in Andhra Pradesh in 2004

Item	*Bt Cotton*	*Non-Bt Cotton*
Casual labour		
Male (in days)	2.61	3.09
Female (in days)	48.80**	38.75
Children (in days)	5.34*	3.46
Total man-days equivalent	37.81**	30.65
Family labour		
Male (in days)	14.71*	12.75
Female (in days)	14.03*	12.36
Children (in days)	0.37	0.36
Total man-days equivalent	24.25**	21.17
Attached labour		
Male (in days)	4.17*	2.98
Female (in days)	0.09	0.22
Children (in days)	-	-
Total man-days equivalent	4.23*	3.13
Total labour utilised		
Male (in days)	21.49*	18.82
Female (in days)	62.92**	51.33
Children (in days)	5.71*	3.82
Total man-days equivalent	66.29**	54.95

Source: Field Surveys.

Sum-up of Bt Cotton Performance

It was found that the Bt cotton technology is superior to the conventional cotton hybrids in terms of yield and net returns in all the agroclimatic zones. The production function analysis brings out clearly that Bt cotton impacts the yield significantly and positively. The Bt farmers from all size categories and all agroclimatic zones benefitted from its cultivation compared to non-Bt farmers from the same categories, though there are differences in the extent to which they benefitted. The study also proved that many of the small farmers participated in using the technology and improved their position with regard to profitability by growing Bt cotton. Above all, the fact that 99 per cent of the sample farmers have grown Bt cotton again in 2006-07, clearly indicates that this technology is useful to the cotton farmers. However, the benefits in rainfed farming are somewhat lower and not statistically significant. It highlights the need to focus on research aimed at abiotic stresses like drought tolerance etc., using the tools of modern biotechnology.

Concluding Observations

While the falling per unit yields in many of the crops since the early 1990s in the country necessitate sustaining the yield levels, closing the yield gaps mainly in the unfavourable areas and small farms, as well as breaking the yield barriers are the urgent tasks in Indian agriculture. Contrary to the situation in the early green revolution days when increasing the yields have been the sole concern, there are many demands now. Some of them are—protecting the environment, minimising the input use including water and catering to the rising food quality and safety concerns. The tools of biotechnology offer a good opportunity in achieving these goals in the coming decades and the continuing pace of discoveries in biology makes it possible to understand the function of genes in various plants so that we can improve and accelerate conventional plant breeding and also modify them to suit the needs. It has also to be recognised that it is not a panacea and development of infrastructure, markets, breeding capacity, input delivery and extension services to reach remote areas are important in enabling a pro-poor agricultural development.

The major difference between the green revolution technologies and the biotechnology is that the locus of research has changed from the public sector to the private sector. In addition, many of these private players are multinational companies from developed countries. Therefore, entire focus of their research is directed towards commercial agriculture in the industrialised countries in view of the market size and profitability. The complementary product development strategies like the herbicide tolerance, which accounts for nearly 70 per cent of the commercialised crop traits, are manifestations of this. The market failures in biotechnology research dominated by these MNCs can be very severe and can potentially neglect 'agriculture common goods'. They may not be interested in crops and traits of importance to the small farmers in developing countries. It is clearly evident from the fact that the ongoing research and trials target neither major food crops in the world like rice and wheat, nor orphan crops like minor millets, pulses and oilseeds. The traits like—increasing the yield potential or tolerance to abiotic stresses like drought of salinity are not aimed at. The proactive role of public sector in undertaking research complementary to that by the private sector is very crucial in this regard. They can also make the technology more affordable by bringing open pollinated varieties as in China. Otherwise, there is a possibility that this technology may bypass the poor.

The empirical study of Bt cotton performance reveals that this technology has the potential to increase the yields by closing the gap between the actual and potential through reducing the biotic pressure from pests. In fact, it is a success story in cotton in the country, where the yield level has risen in India from a five-year average yield level of 204 kgs per hectare of lint at the end of 2002-03 to 470 kgs per hectare in 2007-08.[15] The study also brought out that it can contribute to equity as it helped in significant gains across all agroclimatic zones, size and social categories of farmers. Further, this technology is scale neutral and beneficial to all groups of farmers.

This success can and needs to be replicated for other crops of importance to numerous small farmers of the country by investing in biotechnological research. A rich body of literature is emerging internationally on how to access the discoveries in private biotechnology research in the public domain to develop 'agriculture common goods'.[16] India has some advantages in this regard. A strong National Agricultural Research System (NARS), vibrant private seed market[17] and the public acceptance indicated by rapid adoption rates in cotton attract great deal of research on the crops and traits of relevance in India. The strong NARS and allied research infrastructure also enables us to undertake 'upstream research' to later develop required products. The public sector has to

15. The CACP in its report attributes this significant increase in yield in the past few years mainly to the introduction of Bt technology (GoI, 2008).

16. See for e.g., Byerlee and Fischer (2001), Naylor *et al.* (2004) and Spielman (2007).

17. See Pray *et al.* (2001) and Ramaswami (2002) for detailed analysis.

redefine the priorities in view of these opportunities and formulate a strategy to plan its research as complementary to that in the private sector instead of duplicating the same work. The immediate task is to integrate the tools of biotechnology in a comprehensive agricultural research and development programme. Biotechnologies are not stand alone technologies and have more to do with knowledge created in related fields. Therefore, the collaborative efforts between the agricultural institutes and other research institutes have to be stepped up so that complementarities and economies of scale can be exploited. It is also instructive for the developing countries to push for strengthening the CGIAR network to bring out useful varieties through stepping up basis and applied research in agricultural biotechnology.

The study proved beyond that major share of the gains from products of private research in agricultural biotechnology could go to farmers. Therefore, there is a need to create an enabling environment for the sector to continue and expand research in the country. Strengthening the regulatory framework to properly assess the agronomical, economic, social, environmental and biosafety issues arising from the products of biotechnology *vis-à-vis* conventional products are important to allay the fears of consumers on the one hand and quickly commercialise the products without discouraging the private investment. The state will have to play a proactive role in priority setting for pro-poor agricultural development, increasing investments in basic and applied biotechnology research accessing tools and technologies from private sector through forging creative public-private partnerships to focus on crops and problems of importance to small farmers.

References

A.C. Nielsen (2004). *Nationwide Survey by AC Nielsen ORG-MARG Underscores Benefits of Bollgard™ Cotton, Press Release.* Mumbai: A.C. Nielsen ORG-MARG.

Bhalla, G.S. (2006). *State of the Farming Community in India.* New Delhi: National Book Trust of India.

Birthal, S. Pratap, O.P. Sharma and Sant Kumar (2000). "Economics of Integrated Pest Management: Evidences and Issues", *Indian Journal of Agricultural Economics* 55(4): 644-658. October-December.

Byerlee, Derek and Ken Fischer (2001). "Accessing Modern Science: Policy and Institutional Options for Agricultural Biotechnology in Developing Countries", *World Development* 30(6): 931-948.

Damodaran, A. (1999). "Regulating Transgenic Plants in India: Bio-safety, Plant Variety Protection and Beyond", *Economic and Political Weekly* 34(13).

Dawe, D., R. Robertson, L. Unnevehr (2002). "Golden Rice: What Role could It Play in Alleviation of Vitamin a Deficiency?", *Food Policy* 27: 541-560.

de Janvry, Alain and Elisabeth Sadoulet (2002). "World Poverty and the Role of Agricultural Technology: Direct and Indirect Effects", *The Journal of Development Studies* 38 (4): 1-26.

Desai, B.M. and N.V. Namboodiri (1997). "Determinants of Total Factor Productivity in Indian Agriculture", *Economic and Political Weekly* 32(52). December 27.

Dev, S. Mahendra (2008). *Inclusive Growth in India: Agriculture, Poverty and Human Development*. New Delhi: Oxford University Press.

Dev, S. Mahendra and Ajit Ranade (1998). "Rising Food Prices and Rural Poor: Going Beyond Correlation", *Economic and Political Weekly* 33(39): 2529-2535. September 26.

Fan, Shenggen and Peter Hazell (2000). "Should Developing Countries Invest More in Less-Favoured Areas? An Empirical Analysis of Rural India", *Economic and Political Weekly* 35(17): 1455-1464. April 22.

FAO, (2004). "The State of Food and Agriculture 2003-04: Agricultural Biotechnology-Meeting the Needs of the Poor?", *FAO Agriculture Series* 35. Rome: Food and Agriculture Organisation of the United Nations.

Gandhi, Vasant P., and N.V. Namboodiri (2006). "The Adoption of Bt Cotton in India: Preliminary Results from a Study", *Working Paper* No. 2006-09-04. Ahmedabad: Indian Institute of Management.

GM Science Review Panel (2003). GM Science Review: First Report-an open review of the science relevant to GM crops and food based on the interests and concerns of the public. London, Department of Trade and Industry (also available at *http://www.gmsciencedebate.org.uk/report/default.htm;accessed* March 2004).

GoI (2000). *Reports of the Commission for Agricultural Costs and Prices for the Crops Sown During 1999-2000 Season*. New Delhi: Department of Agriculture and Cooperation, Ministry of Agriculture, Government of India.

———. (2005). *Reports of the Commission for Agricultural Costs and Prices for the Crops Sown During 2004-05*. New Delhi: Department of Agriculture and Cooperation, Ministry of Agriculture, Government of India.

———. (2008). *Reports of the Commission for Agricultural Costs and Prices for the Crops Sown during 2007-2008 Season*. New Delhi: Department of Agriculture and Cooperation, Ministry of Agriculture, Government of India.

Ghosh, Nilabja (2004). "Reducing Dependence on Chemical Fertilisers and its Financial Implications for Farmers in India", *Ecological Economics* 49: 149-162.

Ismael, Y., R. Bennett and S. Morse (2002). "Benefits from Br Cotton Use by Smallholder Farmers in South Africa", *AgBioForum* 5: 1-5.

Herdt, Robert, Gary Toenniessen and John O'Toole (2007). "Plant Biotechnology for Developing Countries", in Robert Evenson and Prabhu Pingali (eds.), *Handbook of Agricultural Economics Vol. 3- Agricultural Development: Farmers, Farm Production and Farm Markets*. Amsterdam: North-Holland, Elsevier.

Huang, Jikun, Rauifa Hu, Ruifa Hu, Qinfang Wang, James Keeley and B. Jose Falck-Zepeda (2002). "Agricultural Biotechnology Development, Policy and Impact in China", *Economic and Political Weekly* 37: 2756-2761. July 6.

James, C. (2008). *Global Status of Commercialiased Biotech/GM Crops in 2005.* International Service for the Acquisition of Agri-Biotech Applications, *http://www.isaaa.org/kc/bin/briefs34/pk/pk.htm*

Johnson, D. Gale (2002). "Biotechnology Issues for Developing Economies", *Economic Development and Cultural Change* 51(1): 1-4. October.

Krishna, P.S. Janaki and G. Pakki Reddy (2005). "Agricultural Biotechnology in Developing Countries: Nature and 'Code' in Meeting the Needs of Resource Poor", *Asian Biotechnology and Development Review* 7(3): 37-54. July. ISSN: 0972-7566.

Krishnakumar, Asha (2003). "A Lesson from the Field", *Frontline* 20(11): 123-126. June 6.

Kalaitzandonakes, G.N (2000). "Agrobiotechnology and Competitiveness", *American Journal of Agricultural Economics* 82(5).

Kalirajan, K.P. and R.T. Shand (1997). "Sources of Output Growth in Indian Agriculture", *Indian Journal of Agricultural Economics* 52(4). October-December.

Kumar, P. and M.W. Rosegrant (1994). "Productivity and Sources of Growth for Rice in India", *Economic and Political Weekly* 29(52). December 31.

Kumar, P. and Mruthyunjaya (1992). "Measurement and Analysis of Total Factor Productivity Growth in Wheat", *Indian Journal of Agricultural Economics* 47(3).

Kumar, Praduman, Surabhi Mittal and Mahabub Hossain (2008). "Agricultural Growth Accounting and Total Factor Productivity in South Asia: A Review and Policy Implications", *Agricultural Economics Research Review* 21(2): 145-172. July-December.

Lipton, M. (2001). "Reviving Global Poverty Reduction: What Role for Genetically Modified Plants?", *Journal of International Development* 15(7): 823-846.

Middendorf, Gerad, Mike Skladany, Elizabeth Ransom and Lawrence Busch (1998). "New Agricultural Biotechnologies: The Struggle for Democratic Choice", *Monthly Review* 50(3).

Naik, Gopal, Matin Qaim, Arjunan Subramanian and David Zilberman (2005). "Bt Cotton Controversy: Some Paradoxes Explained", *Economic and Political Weekly* 40: 1514-1517. April 9.

Narayanamoorthy, A. and S.S. Kalamkar (2006). "Is Bt Cotton Cultivation Economically Viable for Indian Farmers? An Emplical Analysis", *Economic and Political Weekly* 41(26): 2716-2724. June 30.

Naylor, Rosamond L., Walter P. Falcon, Robert M. Goodman, Molly M. Jahn, Theresa Sengooba, Hailu Tefera and Rebecca J. Nelson (2004). "Biotechnology in the Developing World: A Case for Increased Investments in Orphan Crops", *Food Policy* 29: 15-44.

Pingali, Prabhu and Terri Raney (2005). "From the Green Revolution to the Gene Revolution: How will the Poor Fare?", *ESA Working Paper* No. 05-09, Agricultural and Development Economics Division, The Food and Agriculture Organisation of the United Nations, Rome, *www.fao.org/es/esa*

Plucknett, Donal L. (1993). *Science and Agricultural Transformation,* Lecture Series. Washington: International Food Policy Research Institute. September 9.

Pray, Carl E., Bharat Ramaswami and Timothy Kelly (2001). "The Impact of Economic Reforms on R&D by the Indian Seed Industry", *Food Policy* 26: 587-598.

Qaim, Matin (2001). "Transgenic Crops and Developing Countries", *Economic and Political Weekly* 36(32).

———. (2003). "Bt Cotton in India: Field Trial Results and Economic Projections", *World Development* 31(12): 2115-2127.

Qaim, Matin and Alain de Janvry (2005). "Bt Cotton and Pesticide Use in Argentina: Economic and Environmental Effects", *Environment and Development Economics* 10: 179-200.

Qaim, Matin, Arjunan Subramanian, Gopal Naik and David Zilberman (2006). "Adoption of Bt Cotton and Impact Variability: Insights from India", *Review of Agricultural Economics* 28(1): 48-58.

Qaim, Matin and Ira Matuschke (2004). "Economics of GM Crops: An International Perspective with Special Reference to Developing Countries", Paper presented at the International Conference on *Biotechnology and India's Development* organised during November 22-24 at Bangalore by the Institute for Social and Economic Change, Bangalore and Sponsored by the Indian Council of Social Sciences Research, New Delhi and the Indo-Dutch Project on Alternatives in Development, New Delhi.

Qayum, Abdul and Kiran Sakkhari (2003). "Did Bt Cotton Save Farmers in Warangal? A Season Long Impact Study of Bt Cotton in Warangal District of Andhra Pradesh". Hyderabad: Andhra Pradesh Coalition in Defence of Diversity and Deccan Development Society.

———. (2004). "Did Bt Cotton Save Farmers in Warangal? A Season Long Impact Study of Bt Cotton in Andhra Pradesh". Hyderabad: Andhra Pradesh Coalition in Defence of Diversity and Deccan Development Society.

Ramaswami, Bharat (2002). "Understanding the Seed Industry: Contemporary Trends and Analytical Issues", *Indian Journal of Agricultural Economics* 57(3): 415-429. July-September.

Rangarajan, C. (2002). "Biological Interventions and Dry Land Agriculture", Lecture at International Workshop on *Biological Interventions for Dry Land Agriculture-Opportunities and Constraints* at Hyderabad, organised by the Andhra Pradesh Netherlands Biotechnology Programme of the Institute of Public Enterprise, Hyderabad on July 18

Rao, N. Chandrasekhara (2004). "Plant Biotechnology and the Emerging Scenario", *Review of Development and Change* 9(1): 69-92.

Rao, N. Chandrasekhara and S. Mahendra Dev (2009). "Socioeconomic Impact of Bt Cotton in Andhra Pradesh: Evidence from Panel Studies". Hyderabad: Centre for Economic and Social Studies (*mimeo*).

Rao, C.H. Hanumantha (1994). *Agricultural Growth, Rural Poverty and Environmental Degradation in India*. New Delhi: Oxford University Press.

Rao, C.H. Hanumantha and Ashok Gulati (1994). "Indian Agriculture: Emerging Perspectives and Policy Issues", *Economic and Political Weekly:* A-158. December 31.

Reece, David J. and Ejnavarzala Haribabu (2007). "Genes to Feed the World: The Weakest Link?", *Food Policy* 32: 459-479.

Sahai, Suman and Shakeelur Rehman (2003). "Performance of Bt Cotton: Data from First Commercial Crop", *Economic and Political Weekly* 38. July 26.

Scoones, Ian (2002a). "Biotech Science, Biotech Business: Current Challenges and Future Prospects", *Economic and Political Weekly* 37(27).

——. (2002b). "Can Agricultural Biotechnology be Pro-Poor", *IDS Bulletin* 33(4).

Shiva, Vandana, Ashok Emani and Afsar H. Jafri (1999). "Globalisation and Threat to Seed Security: Case of Transgenic Seed Trials in India", *Economic and Political Weekly* 34(10 and 11): 601-613. March 6-13.

Spielman, David J. (2007). "Pro-poor Agricultural Biotechnology: Can the International Research System Deliver the Goods", *Food Policy* 32: 189-204.

Stone, Glenn Davis (2002). "Both Sides Now: Fallacies in the Genetic-Modification Wars, Implications for Developing Countries, and Anthropological Perspectives", *Current Anthropology* 43(4): 611-630. August-October.

UNDP (2001). *Human Development Report 2001: Making New Technologies Work for Human Development.* United Nations Development Programme. New York: Oxford University Press.

World Bank (2007). *World Development Report 2008: Agriculture for Development.* Washington, DC.: The World Bank.

13 Are Crop Forecasts Useful for Policy?

Assessing Alternative Methods

Nilabja Ghosh, Sangeeta Chakravarty and Yogesh Bhatt

With the liberalisation of the markets, the role of the government in agriculture is undergoing radical changes. The system of early estimation of crop production has always aided the government's food management decisions in India. With the government becoming a facilitator of market functioning, the public intelligence system is gaining increasing prominence for enabling critical policy as well as business decisions. The association of crop production performance with inflation makes early forecasts particularly useful for taking timely policy actions for mitigating the effects. Crop forecasting is also valuable for macroeconomic-planning and can potentially serve to combat climate change effects.

Today, an increasing number of private players rely on advance estimates for making informed decisions. Such estimates are drawn from varied available sources but government reports are ideally considered to be among the most reliable. The appearance of private reporters has enlarged the economic intelligence system of the country and enabled cross-validation, but multiple and often contradicting forecasts create confusion and trigger speculation. The government is committed to generating economic, timely and the best possible estimates under the circumstances given that public resources are invested for the purpose. The strength of a methodology is ideally the deciding factor of choosing a particular forecast, so that assessment of the methods also becomes vital to the subject.

In monsoonal India, the unpredictable weather plays a major role in determining agricultural performance while other human and natural factors also make the task of forecasting crop production difficult as well as imprecise. Agricultural production is the consequence of decisions taken by millions of farmers operating across the country in diverse conditions in which the prices they face are often uncertain. Further, the crops vary in their responses to rainfall, temperature and management practices all of which make the production of certain crops highly probabilistic. With the growing importance of decentralised planning in the country, crop forecasts even at subnational level have become valuable. They require more intense and costly efforts than national forecasts and face much more practical difficulties. Yet a meaningful and consistent forecast at the national level can ideally be the sum of the disaggregate forecasts.

Forecasting is a subject that spans various disciplines and methods. Since forecasting is also a costly process, assessment of forecasts must take into account the trade-off that is possible between inaccuracy and cost. This paper is a demonstrative attempt to assess select alternative methods of short-term crop forecasting, using multiple criteria of evaluation. We have considered three select *kharif* crops namely rice, groundnut and cotton covering the broad groups food grains, oilseeds and commercial crops and three standard methods of modelling for making the forecasts. One of this is a simple univariate method that makes the least demand on data and resources. The second method is an extension of the first and accommodates the weather effect only in a multivariate framework. In the third method, a further elaboration is achieved by considering multiple explanatory variables in a standard econometric model where the effects of both economic variables and weather are allowed. Assessment is done for the forecasting ability of the models at the state level using different criteria.

Common Methods of Crop Forecasting in India

A forecast is a statement of what may happen based upon present conditions and observations interpreted in the light of previous experience (Taylor, 1924). Agriculture being a lengthy process, short-term crop forecasts essentially imply predicting what the crop production in the current season would likely to be on the basis of the information that is currently available. The reports on probable crop production can be produced as early as when the crops begin to be sown up to the time it is harvested and each successive estimate will be more informed than the previous one.

India has a traditional system of making three or more reports of in-season estimates of crop production in a year. The method drawn on historic land record systems (Bansil, 2002) ranges over eye estimation, local inspections, field to field enumeration and sampling methods.[1] The eye estimation method till today forms a significant part of the reporting system in India as also in other countries. The system has been reviewed regularly, modified and supplemented in the course of time. The development of methodologies in other disciplines has proved useful for supplementing the regular method. For example, with the revolutionary developments of communication technologies, it was only natural to enhance the crop forecasting capabilities using remote sensing or RS.[2] In recent times methodologies developed in economics, meteorology and space science have been incorporated in a supplementary umbrella scheme named Forecasting of Agriculture output using Space, Agro-Meteorology and Land based observations (FASAL) mainly for the purposes of timeliness and cross-validation. However no method is free from non-statistical inaccuracies (see Ghosh *et al.*, 2008). While the regular *girdawari* method is vulnerable to human and managerial errors and complications, even the RS method has its limitations and in many ways is yet in a developmental stage only. Methods involving agrometeorological parameters and soils dynamics are also in trial. Econometric forecasting depends largely on how well the models are specified, the assumptions that have to be made and the availability of data from past.

---

1. In the states that had a cadastral mapping and a temporary revenue settlement (TS) constituting 86 per cent of the geographical area of the country, reports of village official known as the *patwari* based on their personal inspections and subjective assessments of the conditions of the crops in the initial stages are the inputs for the early estimates. This is the *girdawari* statistical system. In the permanently settled (TS) states accounting for 9 per cent of the area and constituting only the states West Bengal, Orissa and Kerala, a system called the 'Establishment of an Agency for Reporting Agricultural Statistics' (EARAS) that combines sampling techniques with the *girdwari* method is implemented. The remaining states are mostly hilly and forested making up 5 per cent of the India's area where personal assessment of village police (*chowkidars*) is still the practice. Final crop yields are estimated through scientific crop cutting estimate (CCE) surveys that evolved since the colonial times but today the yield estimates are based on well-designed CCEs conducted under the General Crop Estimate Survey (GCES) at the nationwide project help in producing scientific estimates of crop yields.
2. The CAPE project or 'Crop Acreage and Production Estimates' was implemented by the Space Applications Centre or SAC in Ahmedabad.

Assessment of Forecasts

Assessment of forecasts provided by alternative methodologies can be made by considering both the accuracy of the forecasts and the different desirable properties of the methodology used. The accuracy can only be assessed by its performance in retrospect, the most popular method being to compare the forecasts with the actual realised values of past years. Qualitative assessments can be as much relevant as quantitative assessment in adjudging a forecast.

a) *Quantitative accuracy:* The root mean squared error (RMSE), the mean absolute percentage error (MAPE) and the Theil statistics are three common statistics (Weiss and Andersen, 1984; Leuthold, 1975) used to make quantitative comparisons between the forecasted values and the realised ones. This criterion minimises the distance between the forecast and the realised value with little regard to the direction of the movement over time. An overestimation is treated equally as underestimation so long as the absolute difference is under control.

b) *Qualitative accuracy:* It may not be reasonable to reject forecasts simply for their large errors when they serve well in anticipating extreme events or in tracking the course of the actual values especially when provisions have to be made on the basis of anticipated shortage. In reality qualitative accuracy may be a more useful property of crop forecasts from the point of view of the policymaker. Decisions such as negotiations for exports *versus* imports, logistical organisation of procurement and stocking *versus* distribution are based on the quality of the year in terms of its production performance. Qualitative assessment based on the propensity of a method to predict the directions of movements and the turning points correctly or wrongly and the intensity of the inaccuracy has been suggested as criteria for evaluation raising much debate in the literature (Dietrich and Gutierrez, 1973; Naik and Leuthold, 1986).

When choosing a method, it is also useful to ask how costly an error in forecasting is (Hendricks, 1963; DeCanio, 1980; Leuthold, 1970), when the method itself is costly. Other features of the method such as delays, expenditure and information intensity among all of which there are important trade-offs, also add to the cost of the forecast. All this needs to be weighed against the accuracy aspect when making a choice. Some of the desirable properties are listed below.

1. *The earliness of predictability:* Planning, decision-making and provisioning in a timely manner make the lead time of forecasting a significant factor. In many cases, a less accurate forecast received earlier in the production process may be more valuable than a more accurate forecast received too late (Mjelde *et al.*, 1988). Especially, an early warning of a crop failure can go a long way in mitigating the undesirable effects like price rise and agrarian distress through appropriate public policy.[3]

2. *Public expenditure on forecasting:* A more accurate forecast can only be based on a more intense use of information and better processing through scientific methods, coming at the cost of skilled personnel, investment on satellites and computers, mobilisation of the government administrative machinery and availability of abundant and high quality data recorded in the past. On the other hand, methodological improvements, which themselves require investment on research in various subjects can bring down the demand for data and manpower. For the remote and less accessible areas the regular method is too costly so that the easier but less reliable techniques are considered more economic. In fact the mostly costly method of using space science could be a cheaper alternative than land based estimates.

3. *Synergy with other functions:* This is a less discussed but nevertheless a valuable and implicitly acknowledged attribute that also helps to bring down cost in a broader sense. The synergies among various alternative functions are exploited in public programmes. Thus, the government's regular method of forecasting has been embedded with the programme for keeping land records and the statistical database. The *patwaris* end up having a good knowledge of the local realities due to the intensive nature of their work and acquire competency for performing other important rural functions such as relief, electoral and developmental works. The surveys conducted in the GCES to collect the data on yield also

3. It may also be possible to reverse or at least attenuate the potential disaster through mobilising public credit, infrastructure and extension even within the growing period. Satellite imagery and to some extent econometric models based on growing season parameters can to some extent accomplish this. In this case, however, there is a further complexity in which the forecast becomes an explanatory variable for the production and the forecaster actively influences the observed error of forecast.

form part of the general maintenance of an updated statistical database and interestingly also forms an important ingredient for the crop insurance scheme in India. Satellite imagery for crop forecast conducted by the ISRO is integrated with other remote sensing functions of immense strategic and social importance. The remote sensing and agrometeorology are also emerging important for the crop insurance scheme. These synergies also help to harness the specialised expertise required for the work from the existing scientist pools.

4. Integration with policy analysis: Forecasts help to anticipate extreme events and to portend crop failures and frame response policies. Another important attribute of the method of forecasting is described as 'sorting' which enables it to distinguish particular future paths under alternative policy scenarios (Bessler and Kling, 1989). The alternative scenarios can be simulated to analyse policy options on part of the government. Some of these variables being under the control of the policymaker, such models obviously help to measure the effects of such 'forcing' variables, and present the consequences of alternate policy decisions. The methods give valuable clue on how policy can be used by the government to guide the farmers' supply behaviour towards nationally desired directions.

Table 13.1

Advantages and Cost of Three Major Methods of Forecasting

Method	*Advantages*	*Costs*
Regular (*girdwari*, GCE etc.)	Sowing, mid-season final estimates possible, synergy with other administrative work	Manpower shortage, management, statistical skill all leading to poor and delayed estimates
Remote sensing	Mid-season estimates possible, synergy with other remote sensing activities, development of space science capabilities	Sowing season estimate not possible, heavy investment on scientific equipment, high level scientific manpower, method not fully developed in some cases, lack of ground data
Econometric	Earliest (even pre-sowing), mid-season estimates also possible, synergy with policy analysis	Manpower and data management, inadequacy of updated data, need for approximations and projections, accuracy depends on econometrician's perceptions, judgments and skill, quality of past data

Estimation

We estimate the models using annual data from 1975–2005 for relevant crops in the state. The data on crop yield rate (in kilograms per hectare) are taken from area, yield and production of principal crops published by the Government of India (GoI) and updated from government website: *www.agricoop.nic.in,* and all other data are from official publications. On an average three crops are chosen for 11 states wherever they are grown.

Alternative Estimation Methods

The subject of forecasting that evolved in Economics represented a marriage between theory and statistics and has been varied. Structural modelling techniques that measured the relationships among economic variables through the lens of theory saw its heydays after Keynes' General Theory was published in the 1930s, but disenchantment gradually emerged surrounding some of the theoretical underpinnings and empirical realities that. The assumptions on expectation formation were questioned. In general, theories change and forecasting methods too follow but with a lag so that a structural approach to forecasting become tied to the latest 'fashion' in theory (Diebold, 1998) and also runs the risk of producing forecast based on outdated theory. In the 1970s, the Box-Jenkins method of univariate modelling based on the autoregressive and the moving average processes was already revolutionising the literature by providing a sophisticatedly simple and parsimonious method as an alternative. Today the time series based approaches are greatly developed to address different concerns and are widely used in finance and economics.

We first consider a standard structural econometric model in a multivariate framework involving exogeneous variables measuring price of crop, the price of the substitute crops, price of input (fertiliser), irrigation and rainfall. This model required the collection of data from multiple sources and their validation and careful specification of the model to take account of the economic and agronomic realities. We then move to a univariate approach and consider the autoregressive integrated moving average (ARIMA) model. This required data only on crop yield over the same time period and the modelling process was fairly mechanical, based on the usual diagnostics to choose the lag lengths. In its pure form, the

ARIMA model ignores the effect of the current season weather. Rainfall is recognised to be a most important but extraneous effect on agriculture contributing to a major part of yield variations. The importance of information on weather in forecasting exercises has been long established (Venn, 1927; Lave, 1963).[4] Keeping this in mind we extend the simple univariate method to include rainfall treated as an exogeneous variable to estimate a multivariate equation ARIMA. Univariate forecasting of agricultural production was made by Rajaraman and Datta (2003), in which the ARIMA model was fitted with dummies for structural break and pulse dummies in case of one time drops. We used the growing season rainfall as an additional variable with the usual autoregressive (AR) and moving average (MA) terms. The ARIMA model with rainfall provides an opportunity to note the response of a shock to the system coming from a one-time failure of rainfall since this feeds into the future outcomes through the lagged terms.

The modelling process in each of the three cases is followed by forecasting using the estimates provided by the model and updated data on the explanatory variables. In the absence of updated data, projections of the exogeneous variables are used. In particular, the non-availability of ready rainfall data for contemporary and foreseeable future times poses a hindrance for both structural econometric methods and the multivariate ARIMA method with rainfall as an additional variable at the stage of forecasting and the solution is either to provide alternate forecasts for different possible weather conditions, or only for a hypothetically normal rainfall situation.

Econometric Model (ECONM)

Econometric modelling especially in discourses on farm supply responses (Narayana, 1988; Ghosh, 1998) is essentially the estimation of causal relationships that determine the variable of interest keeping in view the importance of both economic and non-economic factors. The

4. For instance, in England the Shaw formula, which was remarkably trustworthy, computed the yield by subtracting from a datum of 39.5 bushels per acre a maximum of 79 lbs. for each inch of rain that had fallen during the weeks of September, October and November and the actual results showed an astonishing agreement with the subsequent output as given in the official returns (Venn, 1927).

modeller's judgment and empirical knowledge are crucial to the identification of explanatory variables after database has been constructed. Some of these variables are known to the farmer at the time of decision-making and others are not and there is intense dependence on how farmers' expectations are postulated. The main advantage of the econometric method is its ability to supply reasonable and informed forecasts at an early date. Further, the model has the power to create alternative scenarios of policy decisions and help policymaking (Bessler and Kling, 1989).

The general model used in this paper is

$$Y_t = a_0 + a_1 P_t + a_2 P^s_t + a_2 I_t + a_3 R_t + a_4 R_t^2 + a_5 D \quad \ldots 1$$

Table 13.2

Summary Details of the Structural Econometric (ECONM) Model

S. No.	*Crop*	*State*	*t-statistics for Price Variable*	*R Square*
1.	Cotton(3)	Andhra Pradesh	2.99 (0.006)	0.26
2.	Cotton(3)	Gujarat	4.44 (0.0001)	0.51
3.	Cotton	Karnataka	3.80 (0.0007)	0.34
4.	Cotton	Maharashtra	2.37 (0.028)	0.58
5.	Cotton(1)	Madhya Pradesh	3.51 (0.002)	0.32
6.	Cotton	Punjab	2.47 (0.021)	0.44
7.	Cotton	Tamil Nadu	1.21 (0.237)	0.29
8.	Groundnut(1)	Andhra Pradesh	1.82 (0.08)	0.29
9.	Groundnut	Gujrat	2.22 (0.035)	0.59
10.	Groundnut(3)	Maharashtra	1.85 (0.075)	0.41
11.	Groundnut	Tamil Nadu	1.98 (0.06)	0.42
12.	Groundnut(1, 3)	Uttar Pradesh	1.97 (0.061)	0.32
13.	Rice(3)	Andhra Pradesh	3.34 (0.003)	0.73
14.	Rice(1)	Bihar	1.15 (0.26)	0.6
15.	Rice(1)	Madhya Pradesh	1.52 (0.142)	0.57
16.	Rice(2)	Orissa	1.86 (0.075)	0.64
17.	Rice(1)(3)	Punjab	2.42 (0.023)	0.7
18.	Rice(3)	Tamil Nadu	5.37 (0.00)	0.59
19	Rice (1)(3)	Uttar Pradesh	5.85 (0.00)	0.88
20	Rice	West Bengal	3.55 (0.002)	0.81

Note: (1) Quadratic term RF^2 is also used.
(2) Price is deflated by wholesale price index of all commodities.
(3) Price is given by revenue/hectare, see Equation 1.
Probabilities given in the parentheses are for the hypothesis of no impact of Price].

in which Y_t is the yield rate, P_t is the price of the crop, P_t^s is the price of substitute crop, I is irrigation, R_t is rainfall in the *kharif* season, R_t^2 represents squared rainfall and D represents dummy as required. Keeping in view the limited database as available from official sources and the non-quantifiability of several effects, the equations are specified appropriately based on the diagnostics (see Appendix A-13). The substitute crops are taken from among major competing crops grown in the state in the *kharif* season. Expectations are based on one period lags so that previous harvest season prices are assumed to be a proxy for the expectation. Since cost of input is a consideration, the output prices are deflated by fertiliser price. Irrigation intensity is another exogeneous variable. Serial correlation is tested using the Breusch-Godfrey Serial Correlation Lagranges Multiplier test and correction is made for autocorrelation when required. Multicollinearity usually leads to high sensitivity of the estimates to the sample at hand. The absence of multicollinearity is ensured as far as possible by dropping up to three observations from the end of sample and comparing the coefficients in the re-estimated equations with the full sample estimates. The stability of the estimates is also thereby considered to indicate their robustness.

The specification chosen for the model is based on the t-statistics and meaningful signs of parameters. In general the variable is retained if its t-statistic exceeds unity. The squared rainfall term is significant only for groundnut in Andhra Pradesh and Uttar Pradesh, for rice in Bihar, Madhya Pradesh, Punjab and Uttar Pradesh and for cotton in Madhya Pradesh. Prices are deflated by wholesale price index of all commodities for rice in Orissa, by prices of substitutes for groundnut in Andhra Pradesh and by fertiliser prices for all other remaining cases based on the diagnostics. Since yield rates have themselves shown dynamics owing to technological advances, economic gain from growing a crop may also be influenced by the expected yield. We have therefore considered both per unit price of product and revenue per hectare as alternative measures of the incentive variable and the final choice depended on the significance of the variable and the R-bar square of the equation. For cotton crop in Maharashtra and Punjab, dummy variable is also introduced after year 2003 because of the introduction of Bt cotton variety with a sharp effect on yield. The irrigation variable is measured by the availability of total irrigation per total hectares of cropped area. The signs of the coefficient are viewed in the light of available theory. Thus, the price effect (reported in Table 13.2) is found to be positive in all cases though insignificant in 3 of the 20 cases. Similarly the sign of the substitute crop price is negative (See Appendix A-13 for details).

Univariate Forecasting (ARIMA)

The use of univariate databased methodology and its variations has rapidly developed context of the financial markets. These methods look at the past stochastic behaviour of the same (dependent) variable (the data speaks for itself) and emphasise the properties of economy. Such methods are not very common for forecasting crop production perhaps due to the excessive economy of information and need for long and generally high-frequency data series. Nevertheless, ARIMA method has been used to forecast crops either in univariate or multivariate forms (Brandt and Bessler, 1982; Schmitz and Watts, 1970; Bessler and Kling, 1989; Rajaraman and Datta, 2003). The main advantage is their relative simplicity and economy in terms of time and data. The role of explanatory variables is minimum or absent and they make no presumption on farmers' notions. The most glaring defect of the method is its overwhelming reliance on past in these models. Extraneous changes with no demonstrated internal dynamics cannot be captured in the forecasting.

An ARMA model is defined in the following form:

$$Y_t = \theta + \alpha_1 Y_{t-1} + \ldots + \alpha_p Y_{t-p} + \mu_t + \beta_1 \mu_{t-1} + \ldots + \beta_q \mu_{t-q}$$

Where p and q are the orders of the series Y_t. If the series is stationary at level the above ARMA (p,q) is estimated. If the series is non-stationary it is differenced and an ARIMA (p,d,q) model is estimated, where d is the order of integration. In the first step, we test whether the yield series are stationary by using the Augmented Dickey Fuller (ADF) test. The variables are differenced accordingly to obtain stationarity. A series is typically represented by the ARIMA (p,d,q) in which it is differenced d times and p is the number of autoregressive terms and q is the number of moving averages The criteria used for selecting the p, q are Akaike Information Criterion (AIC) and Schwartz Bayesian Criterion (SBC) and thereby best-fit model are selected (Table 13.3).

Table 13.3

Summary Details of the ARIMA Model

S.No.	Crop	State	(p,d,q)	AIC	SBC
1.	Cotton	Andhra Pradesh	(1,1,1)	11.11	11.25
2.	Cotton	Gujarat	(0,1,1)	11.86	11.95
3.	Cotton	Karnataka	(1,1,1)	9.57	9.71
4.	Cotton	Maharashtra	(1,1,2)	9.58	9.77
5.	Cotton	Madhya Pradesh	(1,1,1)	9.26	9.40
6.	Cotton	Punjab	(1,1,1)	12.3	12.44
7.	Cotton	Tamil Nadu	(2,1,1)	10.02	10.21
8.	Groundnut	Andhra Pradesh	(0,1,1)	13.36	13.5
9.	Groundnut	Gujarat	(0,1,1)	15.33	15.43
10.	Groundnut	Maharashtra	(0,1,1)	12.75	12.84
11.	Groundnut	Tamil Nadu	(1,1,1)	14.45	14.59
12.	Groundnut	Uttar Pradesh	(1,0,1)	12.99	13.13
13.	Rice	Andhra Pradesh	(1,1,2)	13.13	13.32
14.	Rice	Bihar	(1,0,1)	13.34	13.43
15.	Rice	Gujarat	(1,1,1)	13.56	13.71
16.	Rice	Karnataka	(1,2,2)	13.67	13.86
17.	Rice	Madhya Pradesh	(1,1,1)	13.39	13.53
18.	Rice	Orissa	(0,1,1)	13.73	13.83
19.	Rice	Punjab	(2,1,2)	13.07	13.3
20.	Rice	Tamil Nadu	(2,1,1)	14.54	14.73
21.	Rice	Uttar Pradesh	(1,1,1)	13.38	13.52
22.	Rice	West Bengal	(1,1,1)	12.5	12.64

ARIMA Model with Rainfall (ARIMARF)

The model starts with estimating the ARIMA model as described in the preceding section. An exogeneous variable is then added to the usual terms in the form of rainfall. The rainfall variable is retained only when it has a coefficient with a t-statistic exceeding one. In other cases the resultant selection is same as the ARIMA model. Thus, the ARIMARF model with rainfall is less parsimonious than the pure ARIMA model. Compared to the econometric model that also incorporates economic, infrastructural and policy variables it is significantly simpler. The order of rainfall variable is taken same as the order of the integration. The t-statistics corresponding to the rainfall variable are reported in Table 13.4. Thus, an ARIMARF model is used in the following form:

$$Y_t = \theta + \alpha_1 Y_{t-1} + \ldots + \alpha_p Y_{t-p} + \mu_t + \beta_1 \mu_{t-1} + \ldots + \beta_q \mu_{t-q} + cR_t$$

Where p, q and d are as in the ARIMA model.

Table 13.4

Summary Details of the ARIMA Model with Rainfall (ARIMARF)

S.No.	Crop	State	(p,d,q)	AIC	SBC	Rainfall t-stats	R Square
1.	Cotton	Andhra Pradesh	(1,1,1)	10.54	10.73	1.86 (0.075)	0.61
2.	Cotton	Gujarat	(0,1,1)	11.50	11.64	3.89 (0.0006)	0.30
3.	Cotton	Karnataka	(1,1,1)	10.14	10.33	2.27 (0.032)	0.20
4.	Cotton	Maharashtra	(1,1,2)	9.62	9.85	-0.87 (0.390)	0.46
5.	Cotton	Madhya Pradesh	(1,1,1)	9.24	9.43	1.41 (0.170)	0.27
6.	Cotton	Punjab	(1,1,1)	12.23	12.41	-1.29 (0.210)	0.06
7.	Cotton	Tamil Nadu	(2,1,1)	10.63	10.87	-0.35 (0.73)	0.17
8.	Groundnut	Andhra Pradesh	(1,0,1)	13.26	13.44	2.31 (0.029)	0.20
9.	Groundnut	Gujarat	(0,1,1)	14.24	14.38	7.28 (0.00)	0.85
10.	Groundnut	Maharashtra	(0,1,1)	12.80	12.94	-0.58 (0.565)	0.38
11.	Groundnut	Tamil Nadu	(1,1,1)	14.38	14.57	2.29 (0.031)	0.63
12.	Groundnut	Uttar Pradesh	(1,0,1)	12.97	13.16	1.55 (0.132)	0.17
13.	Rice	Andhra Pradesh	(1,1,2)	13.08	13.31	1.99 (0.058)	0.51
14	Rice	Bihar	(0,1,1)	13.13	13.27	2.91 (0.007)	0.45
15.	Rice	Gujarat	(1,1,1)	13.38	13.57	6.19 (0.00)	0.74
16.	Rice	Karnataka	(1,2,2)	13.32	13.56	4.32 (0.0003)	0.65
17.	Rice	Madhya Pradesh	(1,1,1)	12.82	13.01	4.68 (0.0001)	0.79
18.	Rice	Orissa	(0,1,1)	13.14	13.28	5.012 (0.00)	0.76
19.	Rice	Punjab	(2,1,2)	12.65	12.94	-3.68 (0.001)	0.73
20.	Rice	Tamil Nadu	(2,1,1)	14.57	14.81	1.081 (0.291)	0.05
21.	Rice	Uttar Pradesh	(1,1,1)	12.95	13.14	4.54 (0.0001)	0.62
22.	Rice	West Bengal	(1,1,1)	12.54	12.72	1.61 (0.120)	0.59

Note: Probabilities given in the parentheses are for the hypothesis of no impact of rainfall.

Table 13.5

Rating of Alternative Methods by Different Criteria

Description	*ECONM*	*ARIMARF*	*ARIMA*
Data intensity	1	2	3
Personal involvement/judgment	1	2	3
Lead time advantage	3	2	1
Need for approximation/interpolation	1	2	3
Need for rainfall data	YES	YES	NO
Costliness	1	2	3

Note: 3: least, 2: medium, 1: most.

Rainfall Effect

ARIMARF model is differentiated from the other two methods by its treatment of other variables and most notably the rainfall. In the ECONM model the effect of rainfall is captured by the coefficient of the rainfall variable but a one time shock coming from a poor rainfall affects only the performance in that year. In the ARIMA model the rainfall variable is not separately considered and the effect is in fact subsumed in the disturbance term and the effect of a known shock on rainfall cannot be separately measured. In the ARIMARF model the effect of rainfall can be measured directly on the current period yield rate and also on subsequent years' stream of yield rates, coming down through the autoregressive terms. The literature on agricultural risk has shown ample evidence that the impact of single drought year can linger over a number of subsequent years (Binswanger, 1978 and 1980; Jodha, 1978). Such persistence may be explained by the depletion of productive assets and by the ineligibility for accessing credit. On the contrary, a severe crop loss may also trigger political responses of debt forgiveness, interest waiver and disaster relief all of which may have a temporarily positive impact in the following year. The responses in the future periods may be in either direction depending on the signs of the coefficients and are expected to die down with time. We have plotted the impulse responses from a shock on the growing season rainfall by one standard deviation (Figures 13.1a, 13.1b, 13.1c).

Figure 13.1(a)

Response to a Rainfall Shock in Cotton for Andhra Pradesh

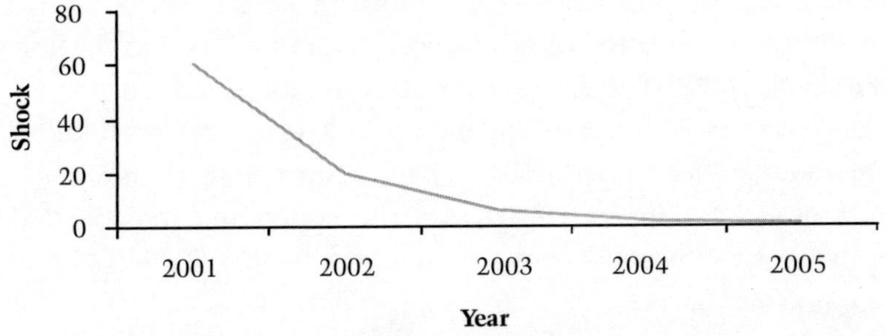

Figure 13.1(b)

Response to a Rainfall Shock in Groundnut for Andhra Pradesh

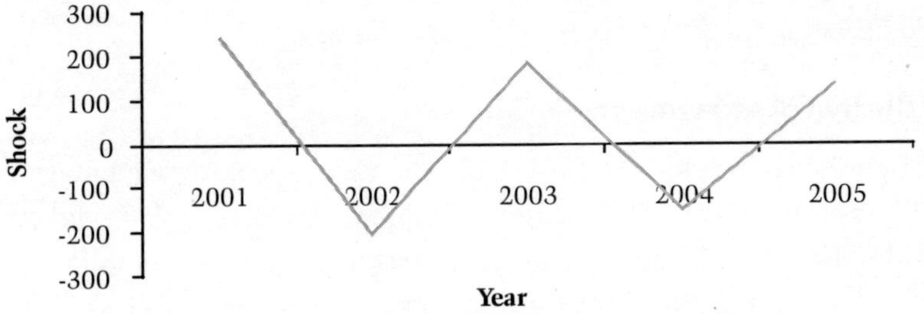

Figure 13.1(c)

Response to a Rainfall Shock in Rice for Andhra Pradesh

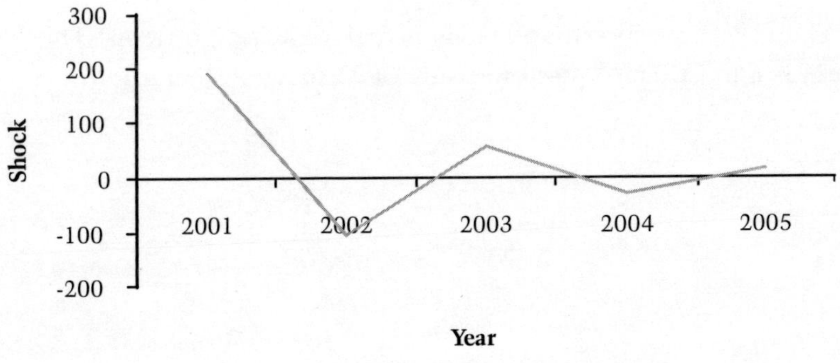

Forecasting

W have used a one-step-ahead forecasting process for the period 2001 through 2005 using a progressively changing sample period. Thus, we estimated each model with the given specification for a period starting with 1975-76 up to 2000-01 to forecast for 2001-02 and so on. In this way we arrive at three sets of forecasts for the forecast years corresponding to the three methods under scrutiny. We try to identify a single method as an umbrella model for forecasting across the states and look for balance among the cost and benefits shown by different criteria (Table 13.5).

Criteria for Assessment

Both quantitative and qualitative accuracy are used as criteria. Thus, we pool the state level forecasts for the five years under four heads: rice, groundnut, cotton and all crops and stack the forecasted and actual values of yield rates for the five forecast years accordingly.

Quantitative Assessment

The following criteria are used: (a) the mean absolute percentage error (MAPE), (b) the THEIL-U1 based on Theil's work (Theil, 1958) and (c) the THEIL-U2 based on Theil's subsequent work (Theil, 1966) as follows:

$$\text{MAPE} = 1/n\sum\left[|P_t - A_t| * 100/A_t\right]$$

$$\text{THEIL-U1} = \sqrt{\Sigma}\ (P_t - A_t)^2 / (\sqrt{\Sigma}\ P_t^2 + \sqrt{\Sigma}\ A_t^2)$$

$$\text{THEIL-U2} = \sqrt{\Sigma}\ (P_t - A_t)^2 / \sqrt{\Sigma}\ A_t^2$$

Where P_t is predicted and A_t is actual yield rate of crop. The Theil measure is a most widely used measure of accuracy in literature.

Table 13.6(a)

Statistics Relating to Forecasting Criteria: THEIL U from Alternative Methods

CROP	*ARIMA*		*ARIMARF*		*ECONM*	
	U1	*U2*	*U1*	*U2*	*U1*	*U2*
Cotton	0.14	0.27	0.12	0.23	0.14	0.26
Groundnut	0.20	0.42	0.15	0.31	0.17	0.35
Rice	0.07	0.14	0.07	0.14	0.23	0.44
Pooled all crops	0.10	0.19	0.08	0.17	0.22	0.43

Note: Here U1 and U2 are as defined by Theil (1958 and 1966), respectively.

Table 13.6(b)

Statistics Relating to Forecasting Criteria: MAPE from Alternative Methods

CROP		*ARIMA*	*ARIMARF*	*ECONM*
Cotton	Mean	22.99	20.87	18.42
	S.D.	13.55	13.05	5.66
	Skewn	1.13	2.21	0.88
Groundnut	Mean	31.30	28.43	24.56
	S.D.	17.71	12.47	14.15
	Skewn	-0.07	-0.61	-0.16
Rice	Mean	14.19	11.71	13.82
	S.D.	8.07	7.11	5.75
	Skewn.	-0.27	0.02	0.20
Pooled all crops	Mean	20.88	18.43	18.12
	S.D.	13.74	12.13	9.12
	Skewn	0.96	1.10	1.02

In Table 13.6(a) we compare among the alternative methods. Both the Theil U1 and U2 statistics criteria show that ARIMARF is the best in cases of groundnut and cotton but in the case of rice ARIMA and ARIMARF perform equally and the rainfall seems redundant for forecasting. Overall, ECONM appears to be the poor. On the contrary the MAPE criteria (Table 13.6(b)) suggests that the ARIMA method performs poorly while ECONM turns out best for cotton and groundnut, though for rice ARIMARF does better. Overall the performance of ECONM and ARIMARF are not much different despite ARIMARF being a less costly method. ECONM estimates show less variance of the error by MAPE criterion. The errors also are mostly positively skewed which also have a favourable connotation that the errors are more likely to be less than the mean values than more. Groundnut is an exception in this regard. Table 13.7 summaries the comparison of the errors through the ratio of the means with respect to the ECONM method.

Table 13.7

Ratio of Mean Error with Respect to Econometric Model

Crop	*MAPE*		*THEIL U2*	
	ARIMA	*ARIMARF*	*ARIMA*	*ARIMARF*
Cotton	1.25	1.13	1.04	0.88
Groundnut	1.27	1.16	1.20	0.89
Rice *kharif*	1.03	0.85	0.32	0.32
Pooled all crops	1.15	1.02	0.44	0.40

Qualitative Forecast Evaluation

A survey of literature suggests the need to look at directional change of the study variable. Kulshreshtha *et al.* (1982) used the ratio of correct directional moves forecasted to total directional moves over the entire forecast period. Theil suggested using the number of over and under predictions and the turning point method to evaluate the qualitative performance. Many researchers like, Leuthold (1969), Meyers *et al.* (1970) and Naik and Leuthold (1986) have used directional change as a measure of qualitative forecast evaluation. Within this class of assessment, the turning point method is used most commonly as a qualitative measure by researchers. A simple turning point method makes use of 2X2 contingency table.

Naik and Leuthold (1986) demonstrate a limitation of turning point method of qualitative evaluation, which can lead to an incorrect interpretation of the results. As an alternative they consider both turning points (peaks and troughs) and upward and downward movements that are not turning points arousing exciting academic debate. One major shortcoming of the turning point method of assessment is that it implicitly requires a forecast of the period even subsequent to the concerned forecast year in order to identify the latter as a peak or a trough.

More often than not a short-term forecaster is interested in knowing how the current yield would compare with the past year rather than with subsequently future year. With an optimistic expectation of the unknown future, one would expect the production performance to improve with time in future. More pertinent questions would be the following: will the production be higher than the previous year's? If there is growth (decline), how large will be the extent of the increase (decrease)? A second question that arises is about the point of comparison. Treating the previous year as the comparison base will not be meaningful if that year is a poor year. Consecutive drought years were not rare in Indian agriculture's history. One way out would be to compare the yield with the moving average of the yield rates of specific number of past years. Even this is not a satisfactory way because even the previous two or three years could be a run of poor years and also because the weights of the remotely past year values will tend to pull down, especially when there is a sharp trend in yield rates. Perhaps, the best years of the past would be a good anchor and a positive movement in this comparison would be a good indication of progressiveness. We have chosen to consider the maximum of the last three years' yield rates as the comparison base Y_b.

$$Y_b = \text{Max}\,(Y_{t-1}, Y_{t-2}, Y_{t-3})$$

To do the assessment we compare both the forecast value and the actually realised value under each method with the past realised value. One advantage of considering the actual values in the comparison base that the data are readily available and we do not lose any information. We have taken two alternate cases.

Growth or No Growth

Under this classification we consider only two cases. The first case (growth) corresponds with one in which the yield rate increases in relation to the base. The classification is made separately for both actual and forecast values and the frequencies (percentage values) are represented in a cross-tabulation of 2X2 contingency table as in Table 13.8(a).

Table 13.8(a)

A 2X2 Contingency Table of Frequencies

		Forecast	
		(↗) growth	(↘) No growth
Actual	(↗) growth	f_{11}	f_{12}
	(↘) No growth	f_{21}	f_{22}

The first entry depicted as f_{11} is the number of cases in which both the actual value and forecast increased in relation to the comparison base and the forecast has also shown an increase. Thus, the diagonal entries f_{11} and f_{22} are the number of cases of correct predictions out of a total number of cases given by the sum of all cell entries and the off-diagonal entries indicate wrong predictions. Assessment is made by the percentage of correct predictions (APR) and that of inaccurate predications (IAPR) given by:

$$APR = (f_{11} + f_{22}) / \Sigma\Sigma f_{ij}$$

$$IAPR = 1 - APR$$

Classified Growth

This method can give more detailed account of predicted and actual direction of yield growth and is more informative. We have distinguished four cases, represented in the 4X4 contingency table (Table 13.8b). Cut-offs or threshold growth rates are required to classify growth performances which could be decided either arbitrarily by specifying a certain rate or more systematically by relating the cut-off rate with the actual observed rate. As a working method we worked out the mean level of the actual positive growth rates and that of actual negative growth rates in the sample to specify the threshold positive growth rate (TPG) and the threshold negative growth rate (TNG). The current approach makes stronger expectation and four cases are defined as follows:

i. High positive growth: Positive growth rate and more than the TPG.

ii. Low positive growth: Growth rate is positive but less than the TPG.
iii. High negative growth: Growth rate is not positive but more than the TNG.
iv. Low negative growth: Growth is negative but less than the TNG.

Table 13.8(b)

A 4X4 Contingency Table of Frequencies

		Forecast			
		(↗) Low positive growth	(↗) High positive growth	(↘) Low negative growth	(↘) High negative growth
Actual	(↗) Low positive growth	f_{11}	f_{12}	f_{13}	f_{14}
	(↗) High positive growth	f_{21}	f_{22}	f_{23}	f_{24}
	(↘) Low negative growth	f_{31}	f_{32}	f_{33}	f_{34}
	(↘) High negative growth	f_{41}	f_{42}	f_{43}	f_{44}

Assessment

In Table 13.9 we find that out of the 110 cases of directional forecasts of the pooled data, there were 39 cases of positive growth with a mean of 18.12 per cent growth rate and there were 71 cases of negative (or no) growth of which the mean was -21.2 per cent. During the five periods for 11 states considered, there were 18 instances of positive growth for rice, 7 for groundnut and 14 for cotton. The mean levels were considered as the threshold levels for classifying the cases of positive and negative growth in the 4×4 case.

Table 13.9

Growth Directions of Crop Yield Movements (Actual)

	Cotton		Groundnut		Rice		Pooled	
	Mean (%)	Numbers	Mean (%)	Numbers	Mean (%)	Numbers	Mean (%)	Numbers
Positive growth	28.18	14	18.52	7	10.13	18	18.12	39
Negative growth	-21.28	21	-31.73	18	-15.2	32	-21.19	71
Total	-1.49	35	-17.6	25	-6.08	50	-7.25	110

In 2×2 case comparing the three methods in 2×2 framework (Table 13.10), ARIMARF has predicted correctly in 71.8 per cent of the cases, followed by ARIMA in 69.1 per cent and then closely by ECONM in 65 per cent. Thus unlike in the quantitative assessment, the ECONM method has not proved superior in a qualitative way. Among the crops, the ARIMARF performed decidedly better than the other two in cases of cotton and rice and in groundnut all the methods perform equally well explaining 60 per cent of the movements correctly.

Table 13.10

Accuracy of Prediction Based on 2X2 Contingency Table (%)

Crop	*Prediction*	*ARIMA*	*ARIMARF*	*ECONM*
Cotton	APR	71.43	74.29	68.57
	IAPR	28.57	25.71	31.43
Groundnut	APR	60.00	60.00	60.00
	IAPR	40.00	40.00	40.00
Rice	APR	72.00	76.00	65.00
	IAPR	28.00	24.00	35.00
Pooled	APR	69.09	71.82	65.00
	IAPR	30.91	28.18	35.00

Note: APR - accurate prediction rate, IAPR - inaccurate prediction rate.

In the 4×4 at the overall level, once again the ARIMARF method performs best predicting correctly in 53.6 per cent cases compared to 43.6 per cent and 50 per cent cases with ARIMA and ECONM methods (Table 13.11).

Table 13.11

Accuracy of Prediction Based on 4X4 Contingency Table (%)

Crop	*Prediction*	*ARIMA*	*ARIMARF*	*ECONM*
Cotton	APR	51.43	60.00	48.57
	IAPR	48.57	40.00	51.43
Groundnut	APR	48.00	44.00	48.00
	IAPR	52.00	56.00	52.00
Rice	APR	36.00	54.00	52.50
	IAPR	64.00	46.00	47.50
Pooled	APR	43.64	53.64	50.00
	IAPR	56.36	46.36	50.00

Note: APR - accurate prediction rate, IAPR - inaccurate prediction rate.

Evaluation

Lack of uniformity of indications is marked in the assessment criteria. Table 13.12 gives a more composite picture of all the exercises conducted in this paper.

Table 13.12

Best Method in Terms of Quantitative and Qualitative Scenario

Crop	*Quantitative*		*Qualitative*	
	MAPE	*THEIL U*	*Growth*	*Classified Growth*
Cotton	ECONM	ARIMARF	ARIMARF	ARIMARF
Groundnut	ECONM	ARIMARF	Draw*	ARIMA, ECONM
Rice	ARIMARF	ARIMA,ARIMARF	ARIMARF	ARIMARF
Pooled	ECONM	ARIMARF	ARIMARF	ARIMARF

Note: * All three methods perform equally. Growth: The criterion is based on binary classification of growth and no-growth.

The ECONM method involving more intense use of a wide range of variables decided on the basis of economic theory stand out in the quantitative assessment only when the MAPE criteria is used. Even in this case the ARIMARF is found best in the case of the crop Rice. The THEIL criterion shows more variety but finds ARIMARF as the best one at the aggregate level. With the qualitative assessment, ARIMARF seems to emerge as the preferable method in all cases. In the simpler and less costly ARIMARF method that combines a univariate time series model with the rainfall variable seems to outperform the others in most ways. Figures 13.2a, 13.2b, 13.2c give a visual representation of the efficiency of the one-period-ahead forecast for the period 2001-02 to 2005-06 pooling by ARIMARF across all the states. Even though ARIMARF appears as a less costly, more timely and relatively accurate method for forecasting, in actual practice rainfall is a major source of indeterminacy at the stage of forecasting. Unless suitable forecasts are provided by the meteorological department, forecast can be made under the assumption of normal rainfall or multiple forecasts made for alternate weather conditions. This assumption becomes an influential determinant of the forecast rather than the unknown reality. However, this weakness affects both ARIMARF and ECONM in the same way.

Figure 13.2(a)

Comparison of Actual and Forecasted Cotton Yield

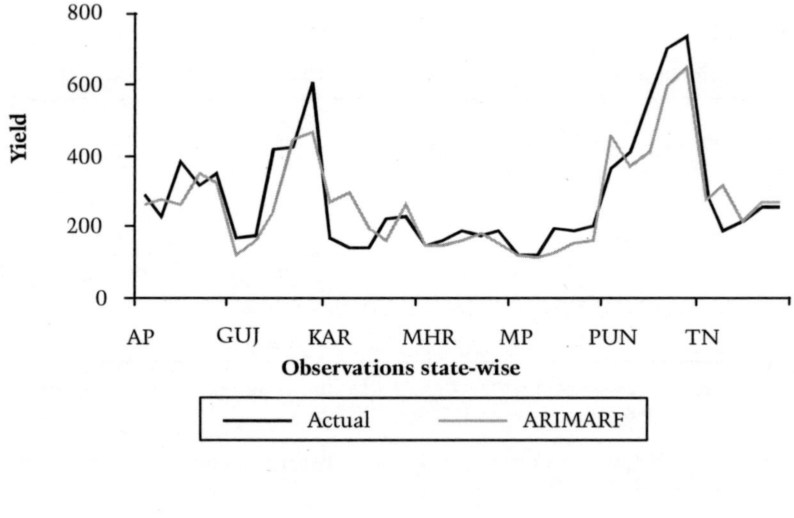

Figure 13.2(b)

Comparison of Actual and Forecasted Groundnut Yield

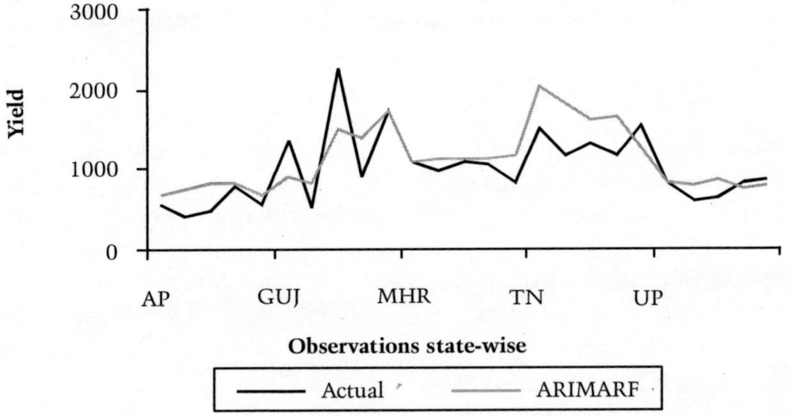

Figure 13.2(c)

Comparison of Actual and Forecasted Rice Kharif Yield

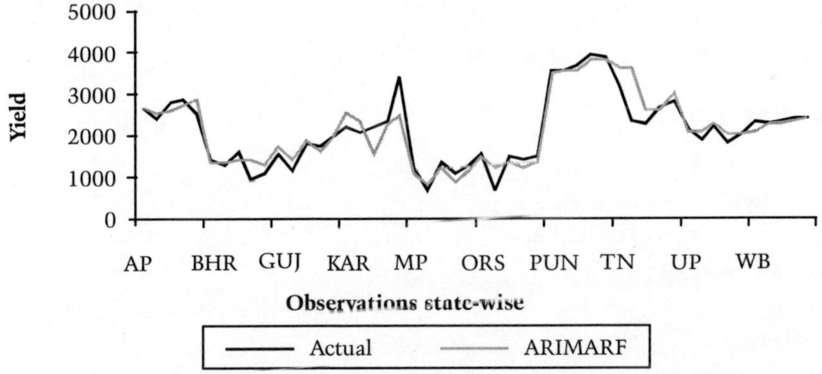

Conclusion

Early crop forecasting is becoming an important function of India's government but as methods of forecasting multiply, comparative assessments need to keep pace. This study takes a cost-benefit perspective to assess three alternative databased methods. The results signify that parsimonious techniques may serve well for forecasting crop yields than data intensive methods provided early forecasts of rainfall are available.

Appendix A-13

1. The Structural Econometric Model

This model is estimated for each crop state-wise. The general form is as follows:

$$Y_t = \alpha + \beta'_j \Sigma X_{it}$$

Where Y is the dependent variable which is the crop yield rate and X_j is the jth explanatory variable, t being the time point. The parameters estimated are α and β. The variables X_j are both economic and non-economic in nature and their choice is made based on economic reasoning. The explanatory variables include price of crop, price of substitute crop, irrigation intensity and rainfall.

Cotton

Maharashtra

CTY =-4.90 + 4.57 Price+51.77 *DUM2003+107.51 Irrigation- 0.041Rainfall
(-0.11) (2.37) (2.95) (2.25) (-1.12)

Adj. R^2 = 0.57

D.W. = 2.13

Price = Price deflated by fertiliser price.

Rainfall = Average rainfall of June to September months.

Groundnut

Andhra Pradesh

GNYLD = -377.12 + 50.15 Price + 10.02 Rainfall - 0.02 Rainfall^2
(-0.68) (1.82) (1.55) (-1.24)

Adj. R^2 = 0.29

D. W. = 1.87

Substitute = *Jowar*

Price = Price deflated by substitute price.

Rainfall = Average rainfall of June to September months.

Gujarat

GNY = -660.55 + 57.80 Price + 1.64Rainfall
(-2.24) (2.22) (6.36)

Adj. R^2 = 0.59

D.W. = 1.76

Price = Price deflated by fertiliser price.

Rainfall = Average rainfall of June to September months.

Uttar Pradesh

$$GNYLD = \underset{(-1.26)}{-677.37} + \underset{(1.97)}{0.02}\ Price + \underset{(2.12)}{15.65}\ Rainfall - \underset{(-1.99)}{0.05}\ Rainfall^2$$

Adj. R^2 = 0.32

D. W. = 2.08

Price = Revenue deflated by fertiliser price.

Rainfall = Average rainfall of June to October months.

Rice

Bihar

$$RY = \underset{(-3.69)}{-2662.70} + \underset{(1.15)}{51.13}\ Price + \underset{(3.26)}{1329.99}\ Irrigation + \underset{(4.15)}{18.53}\ Rainfall - \underset{(-3.86)}{0.03}\ Rainfall^2$$

Adj. R^2 = 0.60

D. W. = 2.00

Price = Price deflated by fertiliser price.

Rainfall = Average rainfall of June to September months.

Madhya Pradesh

$$RY = \underset{(-2.14)}{-1730.34} + \underset{(1.52)}{36.37}\ Price - \underset{(-1.51)}{90.11}\ Substitute\ Price + \underset{(2.18)}{280.79}\ Irrigation + \underset{(2.71)}{4.31}\ Rainfall - \underset{(-2.26)}{0.002}\ Rainfall^2$$

Adj. R^2 = 0.57

D.W. = 2.13

Substitute = Maize.

Price = Price deflated by fertiliser price.

Substitute Price = Price deflated by fertiliser price.

Rainfall = Rainfall of June to September months.

Orissa

$$RY = \underset{(-0.36)}{-234.40} + \underset{(1.86)}{168.90}\ Price - \underset{(-2.32)}{461.12}\ Fertiliser\text{-}Price + \underset{(2.08)}{1163.30}\ Irrigation + \underset{(2.96)}{0.57}\ Rainfall$$

Adj. R^2 = 0.64

D.W. = 1.95

Price = Price deflated by WPIA.

Fertiliser Price = Price deflated by WPIA.

Rainfall = Average rainfall of June to September months.

Punjab

$$RY = 2837.47 + 0.05\ \text{Price} - 179.69\ \text{Irrigation} + 3.84\ \text{Rainfall} - 0.02\ \text{Rainfall}^2$$
(6.63) (2.42) (-2.07) (1.49) (-2.31)

Adj. R^2 = 0.71

D.W. = 1.34

Price = Revenue deflated by fertiliser price.

Rainfall = Average rainfall of June to September months.

Tamil Nadu

$$RY = -932.62 + 0.07\ \text{Price} + 5793.82\ \text{Irrigation} + 2.83\ \text{Rainfall}$$
(-1.0) (5.37) (2.81) (1.10)

Adj. R^2 = 0.59

D.W. = 1.13

Price = Revenue deflated by fertiliser price.

Rainfall = Rainfall of July month.

West Bengal

$$RY = -546.55 + 245.82\ \text{Price} + 1865.99\ \text{Irrigation} + 0.90\ \text{Rainfall}$$
(-1.89) (3.55) (5.18) (1.16)

Adj. R^2 = 0.80

D.W. = 1.38

Price = Price deflated by fertiliser price.

Rainfall = Average rainfall of June to September months.

ARIMA and ARIMARF Model

An ARMA model is defined in the following form

$$Y_t = \theta + \alpha_1 Y_{t-1} + \ldots + \alpha_p Y_{t-p} + \mu_t + \beta_1 \mu_{t-1} + \ldots + \beta_q \mu_{t-q}$$

Where p and q are the orders of the series Y_t. If the series is stationary at level, the above ARMA (p,q) is estimated. If the series is non-stationary, it is differenced, ARIMA (p,d,q) is estimated, where d is the order of integration.

The ARIMARF (ARIMA with rainfall) is estimated using the above specification along with the rainfall variable wherever the t-statistic is greater than one. In other cases the above specifications are retained. The rainfall is also differenced, accordingly.

2. *Chi Square Test*

Table A-13.1

Chi Square Test Based On 2x2 and 4x4 Contingency Tables

Crop	*Chi Sq 2x2 Case*	*Asym. Signif. (2-Sided)*	*Chi Sq 4x4 Case*	*Asym. Signif. (2-Sided)*
Actual *vs* ARIMA				
Cotton	5.25	0.02	12.34	0.06
Groundnut	0.20*	0.66	18.66	0.03
Rice	6.45	0.01	17.95	0.04
Pooled	9.64	0.00	23.67	0.01
Actual *vs* ARIMARF				
Cotton	7.20	0.01	23.87	0.01
Groundnut	0.20*	0.66	28.28	0.00
Rice	10.95	0.00	32.43	0.00
Pooled	14.89	0.00	68.28	0.00
Actual *vs* ECONM				
Cotton	3.60	0.06	17.72	0.04
Groundnut	0.00*	0.97	18.52	0.03
Rice	3.13	0.08	19.88	0.00
Pooled	4.08	0.04	46.90	0.00

Note: * Hypothesis is accepted.

References

Askari, H. and J.H. Cummings (1977). "Estimating Agricultural Supply Response with Nerlove Model: A Survey", *International Economic Review*18(2), June.

Bansil, P.C. (2002). *Agricultural Statistics in India*. CBS Publishers.

Bessler, Davis A. and John L. Kling (1989). "Forecast and Policy Analysis", *American Journal of Agricultural Economics* 71(2).

Binswanger, Hans P. (1978). "Risk Attitudes of Rural Households in Semi-arid Tropical India", *Economic and Political Weekly*. June.

———. (1980). "Attitudes towards Risk: Experimental Measurement in Rural India", *American Agricultural Economics Association*. August.

Brandt, Jon A. and David A. Bessler (1982). "Forecasting with Dynamic Regression Model: A Heuristic Approach", *North Central Journal of Agricultural Economics* 4: 27-33.

DeCanio, Stephen J. (1980). "Economic Losses from Forecasting Error in Agriculture", *Journal of Political Economy* 88(2).

Deibold, X. Francis, Lee E. Ohanian and Jeremy Berkowitz, (1998). "Dynamic Equilibrium Economies: A Framework for Comparing Models and Data", *The Review of Economic Studies* 65(3): 433-51 (July).

Dietrich, Kimball J. and Alfredo D. Gutierrez (1973). "An Evaluation of Short Term Forescast of Coffee and Cocoa", *American Journal of Agricultural Economics* 55(1).

Ghosh, Nilabja (1998). *Studies of Supply Responses in Indian Agriculture*. PhD Thesis submitted to Indian Statistical Institute, Calcutta.

———. (2008). "Measuring the Efficacy of Targeted Public Programmes: Public Works Programmes in India", in Basudeb Guha-Khasnobis, S.S. Acharya and B. Davis (eds.), Food *Security Indicators, Measurement, and the Impact of Trade Openness* (by ICSSR, UNU-WIDER and FAO). Oxford University Press.

Ghosh, Nilabja, Sangeeta Chakravarty and Yogesh Bhatt (2008). "Crop Supply Forecasting for Food Management: A Review of Practices, Developments and Options", *Agricultural Situation in India* (Special Number—Theme: Food Management in Global Perspective, August). New Delhi: Ministry of Agriculture, Government of India.

Hendricks, Walter A. (1963). "Forecasting Yield with Objective Measurements", *Journal of Farm Economics* 45(5).

Jodha, N.S. (1978). "Effectiveness of Farmers' adjustment to Risk", *Economic and Political Weekly*. June 1.

Krishna, R. (1963). "Farm Supply Response in India-Pakistan: A Case Study of the Punjab Region", *Economic Journal* 73.

Kulshreshtha, S.N., J.D. Spriggs and A. Akinfemiwa (1982). "A Comparison of Alternative Approachs to Forecasting Cattle Prices in Canada", *University of Saskatchewan Tech. Bulletin* No. 82-01.

Lave, Lester B. (1963). "The Value of Better Information to the Raisin Industry", *Econometrica* 31: 151-164.

Leuthold, Raymond M. (1969). "An Analysis of Daily Fluctuation in the Hog Economy", *American Journal of Agricultural Economics* 51.

———. (1970). "Forecasting Daily Hog Prices and Quantitites: A Study of Alternative Forecasting Techniques", *Journal of American Statistical Association* 65(329).

———. (1975) "On the Use of Theil's Inequality Coefficients", *American Journal of Agricultural Economics* 57(2).

Meyers, L.H., J. Havlicek and P.L. Henderson (1970). *Short-term Price Structure of the Hog Pork*.

Mjelde, James W., Steven T. Sonka, Bruce L. Dixon and Peter J. Lamb (1988). "Valuing Forecast Characteristic in a Dynamic Agricultural Production System", *American Journal of Agricultural Economics* 70(3).

Naik, G. and Raymond M. Leuthold (1986). "A Note on Qualitative Forecast evaluation", *American Journal of Agricultural Economics* 68(3).

Narayana, N.S.S. and K.S. Pari (1981). "Supply Response and Acreage Allocation: A Case of Indian Agriculture", *International Institute for Applied Systems Analysis* RR-81-1.

Naryana, N.S.S. (1988). "Production Decision and Farm Supply Response to Policy Changes: A Review of Theoretical Issues and Evidence from India", *Working Paper*, Indira Gandhi Institute of Development Research, Bombay.

Nerlove, M. (1958). *The Dynamics of Supply: Estimation of Farm supply Response to Price.* Baltimore: Johns Hopkins University Press.

Rajaraman, Indira and Arindam Datta (2003). "Univariate Forecasting of State Level Agricultural Production", *Economic and Political Weekly*, May 03.

Schmitz, Andrew and Donald G. Watts (1970). "Forecasting Wheat Yield: An Application of Parametric Time series Modeling", *American Journal of Agricultural Economics* 52(2).

Taylor, Henry C. (1924). "Agricultural Forecasting", *Journal of Farm Economics* 6(2).

Theil, H. (1958). *Economic Forecast and Policy.* North-Holland Publishing Co.

———. (1966). *Applied Economic Forecasting.* Chicago Rand-McNally and Co.

Venn, J.A. (1927). "Crop Forecasting in England", *The Economic Journal* 37(147): 404-410.

Weiss, A.A. and A.P. Andersen (1984). "Estimating Time Scrics Models Using the Relevant Forecast Evaluation Criterion", *Journal of the Royal Statistical Society* 147(3): 8014.